A Guide to Emergency Lighting

A Guide to Emergency Lighting

Chris Watts

First published in the UK in 2006

by
BSI
389 Chiswick High Road
London W4 4AL

Reprinted 2007

© British Standards Institution 2006

All rights reserved. Except as permitted under the *Copyright, Designs and Patents Act 1988*, no part of this publication may be reproduced, stored in a retrieval system or transmitted in any form or by any means – electronic, photocopying, recording or otherwise – without prior permission in writing from the publisher.

Whilst every care has been taken in developing and compiling this publication, BSI accepts no liability for any loss or damage caused, arising directly or indirectly in connection with reliance on its contents except to the extent that such liability may not be excluded in law.

The right of Chris Watts to be identified as the author of this Work has been asserted by him in accordance with sections 77 and 78 of the *Copyright, Designs and Patents Act 1988*.

Typeset in Century Schoolbook by Monolith
Printed in Great Britain by MPG Books Ltd, Bodmin, Cornwall

British Library Cataloguing in Publication Data
A catalogue record for this book is available from the British Library

ISBN 978-0-580-47755-3

Acknowledgements

The author would like to thank all his colleagues in the UK, Europe and further afield, whose helpful discussions and support have enabled us to produce the present generation of the emergency lighting standards that are now available. The members of the trade associations, Industry Committee for Emergency Lighting (ICEL) and British Fire Protection Systems Association (BFPSA), have been particularly helpful in providing information on the application techniques.

He would also like to record his thanks to Cooper Lighting and Security Limited who have provided information on their product ranges and the performance data used in the production of this book.

Contents

Acknowledgements		v
1.	**Introduction**	**1**
	Emergency lighting terminology	1
	Escape lighting systems	2
	Development of legislation	3
	Legislative requirements	5
	Other UK legislative requirements	7
	European Directives and Recommendations	8
	Fire safety legislation	10
2.	**Emergency lighting standards: an overview**	**19**
	Structure of standards production and administration	19
	Emergency lighting standards	21
3.	**Code of practice for the emergency lighting of premises (BS 5266-1:2005)**	**27**
	Introduction	27
	Changes to the standard	28
	Consultation (clause 4.1)	29
	Vision and visibility (clause 6)	31
	Minimum illuminance and adaptation	32

	Construction of escape lighting luminaires and labelling of categories	34
	Wiring and installation practices	38
	Test facilities	40
	Voltage compatibility of a slave luminaire and a central battery system	41
	Guidance on the category of system to be adopted for typical premises	42
	Design of system	45
	Routine inspections and tests	47
4.	**Emergency lighting (BS 5266-7/BS EN 1838)**	**49**
	Introduction	49
	General principles	49
	Escape routes	51
	Ratio of maximum to minimum illuminance	51
	Compliance with these requirements	53
	Open areas	53
	High risk task area lighting	55
	Safety signs	56
5.	**Application standard (BS EN 50172/BS 5266-8)**	**61**
	General guidance (clause 4.1)	61
	Identification and illumination of emergency exit signs (clause 4.2)	62
	Open area (antipanic) (clause 4.4) (also covered in clause 3.4)	62
	System design (clause 5)	63
6.	**Emergency lighting luminaires (BS EN 60598-2-22)**	**69**
	Self-contained luminaires	70
	Centrally supplied luminaires	70

Contents

7.	**Centrally powered supply systems (battery systems) (BS EN 50171)**	**77**
	Types of central power supply systems (clause 4)	77
	Automatic testing of central power supply systems (clause 6.11)	83
	Batteries (clause 6.12)	84
8.	**Automatic test systems for emergency lighting (IEC 62034 and other standards)**	**87**
	Self-contained with stand alone facilities (type S)	87
	Self-contained with remote panel (type P or EN)	88
	Central powered system with remote panel (type P or EN)	88
	General requirements (clause 4.1)	88
	Timing circuit (clause 4.2)	89
	Test function (clause 4.3.1)	89
	Emergency supply (clauses 4.3.2–4.3.4)	89
	Protection against component and intercommunication faults (clause 4.4)	90
	Testing of lamps (clause 4.5)	90
	Selection of a suitable system	94
9.	**Other relevant standards**	**97**
	BS 5266-2	97
	BS 5266-4	97
	BS 5266-5	98
	BS 5266-6	98
	Battery standards	98
	Luminaire standards	99
	Lighting terms and photometry	100
	Wiring systems	100
	Light and lighting mains lighting for indoor workplaces (BS EN 12464)	100

Light and lighting: sports lighting (BS EN 12193)	101
Emergency lighting (CIE 5-19)	101
Other relevant documents	101
10. Regulatory Reform (Fire Safety) Order	**103**
Legislative background	103
11. The Building Regulations	**113**
Approved Document B	113
Emergency lighting compliance checklist	114
12. System design	**121**
Design objectives	121
Initial considerations	121
Legislative requirements	123
Other UK legislative requirements	125
Predesign information	125
Design of new installations	128
Spacing tables	137
High risk task area lighting	139
Design control procedures	140
Testing and log-book	140
BS 5266-1:1999 test regime	140
Test records	141
13. System selection	**145**
System requirement	145
Power source	146
Typical applications	149
Modes of operation	149

Contents

14. Photometry for emergency lighting — 159
Photometric theory — 159
Lighting requirements — 161
Verification of photometric design — 164
Products that are difficult to provide with photometric data — 165

15. Design considerations for major applications — 167
Hospitals and nursing homes — 167
Hotels and boarding houses — 170
Non-residential premises used for recreation — 172
Shops and covered shopping precincts — 174
General industrial premises and warehouses — 176
Offices — 178
Schools and colleges — 179
Transport locations — 181

16. Installation, maintenance and testing of emergency lighting — 185
Initial procedures — 185
Self-contained systems — 185
Central battery systems — 186

Annexes — 189
Annex A Inspection and test certificates — 191
Annex B Completion certificates — 195
Annex C Compliance checklist form inspection engineers — 203
Annex D Mathematical table for use in photometric calculations — 205
Annex E Legislation, standards and training affecting emergency lighting — 207

1. Introduction

Emergency lighting has been developed over many years to provide illumination to allow occupants to use escape routes in the event of a failure of the normal lighting supply. It has traditionally been associated with fire protection systems and it has become an essential element to enable persons to escape from fires. But it also has an important function in the event of total or local supply failures, protecting users and giving them confidence to escape safely.

Because our task is only to assist people to evacuate buildings we are able to use much lower light levels than are used for normal lighting. However, this makes it essential to design the system correctly to make effective use of it.

Many standards have been produced to endorse established good engineering practice in this important area of life safety. With the changes in requirements and responsibilities defined in current legislation, this book is intended to reveal the intention behind the requirements, to enable them to be interpreted better for specific applications.

Emergency lighting terminology

For the purposes of the British and European standard BS EN 1838, 'Emergency lighting' is the generic term for equipment which provides illumination in the event of failure of supply to normal lighting. There are a number of specific forms, as shown in Figure 1.

Emergency escape lighting

This is defined as that part of emergency lighting that is provided to enable safe exit in the event of failure of the normal supply. (This type of emergency lighting forms part of the fire protection of a building.)

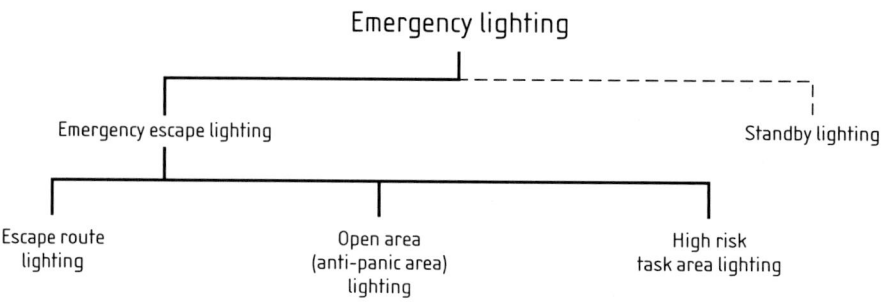

Figure 1 Specific forms of emergency lighting

Standby lighting

This is that part of the emergency lighting provided to enable normal activities to continue in the event of failure of the normal mains supply. (This lighting does not provide fire protection unless it meets the same equipment, design and installation requirements as emergency escape lighting systems.)

Escape lighting systems

Escape route lighting

This is the part of emergency lighting that is provided to enable safe exit for the building's occupants, by providing appropriate visual conditions and direction finding on escape routes and in special areas/locations (e.g. corridors and stairs). It also ensures that fire fighting and safety equipment can be readily located and used.

Open area (or antipanic area) lighting

This is the part of emergency escape lighting that is provided to reduce the likelihood of panic and to enable the safe movement of the occupants towards escape routes by providing appropriate visual conditions and direction finding (e.g. in large rooms).

Introduction

High risk task area lighting

This is the part of emergency lighting that is provided to ensure the safety of people involved in a potentially dangerous process or situation and to enable proper shut down procedures to be carried out for the safety of other occupants of the premises (e.g. to protect persons from dangerous machinery).

Development of legislation

Emergency lighting is demanded mainly because of legislation. This legislation itself is a direct result of public and political pressure to safeguard life and to ensure uniform standards in places where the public gather.

In addition, local legislation is drawn up to meet specific and often local risks where these are not adequately covered nationally. It is the fire authority, jointly with the local authority building inspectors, that enforces legislation and provides a local control.

Legislation is not the only factor in the decision to install emergency lighting which may also be required to meet demands for safety, security, financial protection, or to maintain operation of the site.

History of the legislative process

At one time, the only places which were required by law to install emergency lighting were cinemas, theatres and passenger ships, although emergency lighting was also common in hospital operating theatres. It was also used in some department stores and in premises licensed by the Magistrates' Courts.

However, social conscience was triggered by a few tragic incidents in places of public entertainment, hotels, old people's homes and licensed clubs. This initiated a multitude of mainly disconnected acts, by-laws, regulations and standards. All imply or demand adequate lighting at all material times to provide means of escape for the public and employees.

Some local authorities when faced with large and concentrated risks, and in the absence of adequate national standards and legislation, pushed ahead with their own legislation, regulations and standards.

Unfortunately, they did not agree on the requirements that should be met. This made product and application standardization difficult, and resulted in confusion as to which products were acceptable.

The Factories Act 1961 and the Offices, Shops and Railway Premises Act 1963 were introduced. These included measures to enforce public safety including requiring emergency lighting, in places of work.

These two Acts were administered and enforced at local level by inspectors who were independent of the fire brigades and who advised on fire and emergency lighting equipment in places of public entertainment.

The Fire Precautions Act 1971 and the Health and Safety at Work Act 1974, backed by existing legislation for cinemas, were strenuous efforts to bring order and consistency of requirements. They combined the enforcing powers and administration at local level and put them in the hands of the local authority fire service.

The local authority reorganization of 1975, which created larger fire authority areas and a more uniform structure, improved the effectiveness of the fire authorities as an enforcing power.

The introduction of the British Standard Code of Practice BS 5266-1 in July 1975 further improved the standard and uniformity of emergency lighting installations by defining the illumination needed, and the introduction of the British Standard for Emergency Lighting Luminaires (now developed and numbered as BS EN 60598-2-22) helped to ensure an improved standard of performance and reliability.

National legislation and standards

Legislation is produced when the government responds to the concerns, expressed by public or private bodies, for the need to introduce specific legal controls for the safety of the citizens. The government approaches standards institutes, manufacturers' associations, professional bodies, trade associations, research stations and representatives of those who will be affected to actively help in the drafting of the proposals.

Similarly, when the British Standards Institution (BSI) or any similar organization is producing a code of practice or equipment standard, the same bodies are involved and contribute to and influence the final form of the standard.

Introduction

Legislative requirements

There is a considerable amount of British and European legislation affecting emergency lighting. The major items are given below.

The Construction Products Directive (89/106)

Section 4.3.8.1 of this directive defines emergency lighting installation (panic lighting, escape lighting) as follows:

'The purpose of the installation is to ensure that lighting is provided promptly, automatically and for a suitable time in a specific area when normal power supply to the lighting fails. The purpose of the installation is to ensure that:

- the means of escape can be safely and effectively used;
- activities in particularly hazardous workplaces can be safely terminated;
- emergency actions can be effectively carried out at appropriate locations in the workplace.'

In the UK this directive is implemented by the building control officers and applies to most new and refurbished buildings except for private dwellings.

Details of the requirements are given in part B 1 section 6.36 of the Building Regulations, which specifies that all escape routes and areas listed in Table 9 of this section should have emergency lighting conforming to BS 5266-1.

The 2000 edition of the Building Regulations has been upgraded to require any open areas larger than 60 m^2 in shop, commercial, industrial, storage and other non-residential premises to have emergency lighting (previously this requirement just applied to offices). School buildings without natural light or used outside normal hours must now have emergency lighting.

The Workplace Directive (89/654): Regulatory Reform (Fire Safety) Order

This directive protects the occupants of premises and has two clauses that specifically relate to emergency lighting.

- Clause 4.5: Specific emergency routes and exits must be indicated by signs in accordance with the national regulations.
- Clause 4.7: Emergency routes and exits requiring illumination must be provided with emergency lighting of adequate intensity in case the lighting fails.

In the UK this is implemented by the new guidance document issued by the Office of the Deputy Prime Minister (ODPM) which clarifies that this is done by the user performing a risk assessment for all premises in which people are employed. The fire and rescue authorities are responsible for auditing compliance.

If more than five people are employed there must be a written record of the assessment's findings and the action taken.

If a fire certificate has been issued recently a risk assessment is still required but it is likely that few if any additional fire precautions will be needed. If the fire certificate was given according to an out-of-date standard this must be addressed in the risk assessment.

In the UK the workplace directive is implemented by the Regulatory Reform (Fire Safety) Order which amends or replaces 118 pieces of legislation, the most significant being the repeal of the Fire Precautions Act 1971 and the revocation of the Fire Precautions (Workplace) Regulation 1997. Anyone familiar with the 1997 Regulations will recognize much that is in the Order; it develops and extends many of their concepts.

The Order applies to the majority of premises and workplaces in the UK. But it excludes: dwellings, the underground parts of mines, anything that floats, flies or runs on wheels, offshore installations, building sites and military establishments.

The Order firmly places responsibility on the employer or operator (who may delegate responsibility to a 'responsible person') of the building and outlines the measures that must be taken to ensure the safety of all the people that he or she is directly or indirectly responsible for. At the same time it allows the enforcing authority to make sure that it is enacted (by force if necessary) and defines the penalties the courts can impose in the event of non-compliance.

It requires the responsible person to carry out a fire risk assessment, produce a policy, develop procedures (particularly with regard to evacuation), provide staff training and carry out fire drills. The responsible person must also provide and maintain: clear means of escape, emergency lighting, signs, fire detection, alarms and extinguishers.

Introduction

The Signs Directive (90/664) implemented in the UK by Statutory Instrument 341

The Signs Directive is applicable to fire exit and first aid signs. It defines the format of the signs and specifies that they must be visible at all times that the building is occupied. It has the following clauses which relate to the provision of emergency lighting to enable the signs to be seen in the event of a supply failure.

- Clause 6: Depending on requirements, signs and signalling devices must be regularly cleaned, maintained, checked, repaired, and replaced.
- Clause 8: Signs requiring some form of power must be provided with a guaranteed supply.

In the UK the Health and Safety Executive have passed responsibility for ensuring compliance for fire safety signs to the fire authority. They have produced a combined guidance document covering the use of safety signs.

Other UK legislative requirements

Some workplaces require a licence from the local authority. The fire authority may require higher levels for premises including those which provide:

- the sale of alcohol;
- music and dancing;
- theatres and cinemas;
- gambling;
- sports stadia;
- public entertainment.

Some premises must be registered with the local authority and need to be accepted by the fire authority. These include:

- nursing homes;
- children's homes;

- residential care homes;
- independent schools.

European Directives and Recommendations

The following directives have been adopted into UK legislation and are the subject of guidance documents produced by the relevant ministries.

- Workplace Directive (89/654 EEC)
- Construction Products Directive (89/106 EEC)
- Safety Signs Directive (92/58 EEC)
- Fire Safety in Hotels Recommendation: Requirements for Europe (86/666 EEC)

The Workplace Directive (89/654 EEC)

This is partially implemented in the UK by The Workplace (Health, Safety and Welfare) Regulations 1992. It includes within its scope of premises most buildings where people are employed. The major UK legislation implementing it is now the Regulatory Reform (Fire Safety) Order 2005.

The Regulatory Reform (Fire Safety) Order (2005)

This Order applies to every workplace with certain exceptions such as: ships, construction sites, mines, temporary workplaces, fields, woods or other agricultural or forestry land, aircraft, locomotive or rolling stock, trailers and some vehicles. The Regulations require a risk assessment and an emergency plan to be prepared. The supporting guidance stresses the need for cost-benefit analysis and minimizing burdens commensurate with saving lives and the safe evacuation of premises.

The Workplace Directive is retrospective, i.e. it requires that, over time, all places of work (with the above exemptions) are brought up to standard.

In June 2005 the Fire Precautions Act, together with a number of other smaller fire safety laws, was superseded and replaced by the Regulatory Reform (Fire Safety) Order which enforces the responsible operators of buildings to perform risk assessments and limit the risks to tolerable levels. Compliance with this procedure will be audited by the

Introduction

fire authorities. Details of this legislation and its impact are discussed in Chapter 4.

The Construction Products Directive (89/106 EEC)

This covers both buildings and civil engineering works including domestic, commercial, industrial, agricultural, educational and recreational buildings as well as roads and highways, bridges, docks and tunnels. It requires that such buildings or works are designed and built in such a way that they do not present unacceptable risks of accidents in service or in operation such as stumbling or tripping in poor visibility, and that the safety of occupants and rescue workers is ensured in the case of fire. Minimum standards of illumination are required so that people may move safely within the works, including if they have to escape. In addition, escape routes are required to provide secure and adequate lighting, capable of operating despite failure of the electrical supply.

The Safety Signs Directive (92/58 EEC)

This is retrospective and was implemented in the UK on 1 April 1996. It calls for the provision of emergency signs in all places of work. These signs must be regularly cleaned, tested and maintained, and visible at all times. The traditional text EXIT signs must have been replaced by the pictogram by December 1998. A guide to Statutory Instrument No. 341, The Health and Safety (Safety Signs and Signals) Regulations 1996, has been published by the Health and Safety Executive.

The Fire Safety in Hotels Recommendation

This applies to all establishments with 20 or more paying guests. The Recommendation is intended to reduce the risk of fire breaking out, prevent the spread of flames and smoke, and ensure that all occupants can be evacuated safely. In particular it requires that escape routes and doors are indicated by safety signs visible day and night, and that an emergency lighting system is provided with sufficient duration to enable evacuation for all occupants. Compliance with this recommendation will be covered by the Regulatory Reform (Fire Safety) Order in the UK.

NOTE: the latest edition of documents (directives, standards, guidance notes etc.) should be referred to.

Fire safety legislation

The changes in the way legislation is implemented have resulted in considerable changes in responsibility for the systems. Figure 2 shows that the fire authority defined the areas that needed illumination as provided by compliance with BS 5266-1.
Figure 3 shows that this has changed. The employer is now responsible for assessing the risks in his or her premises and then providing emergency lighting to reduce those risks to tolerable levels. Fire authorities will be able to audit the provisions. So, the employer uses compliance with BS 5266-1 as a way of demonstrating compliance with his or her legal responsibilities.

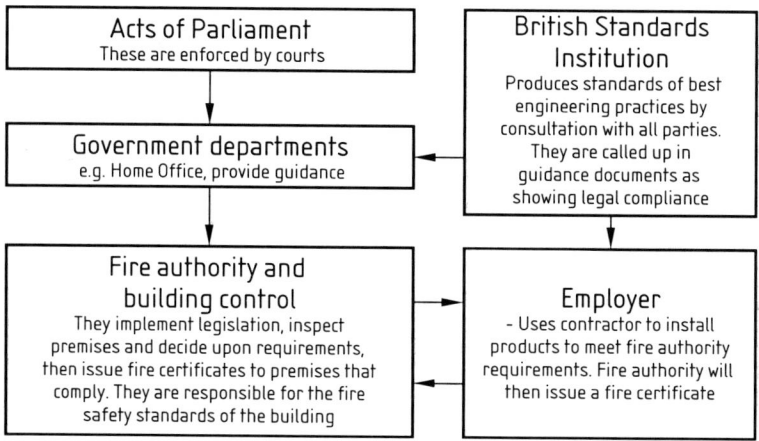

Figure 2 Flowchart of implementation of the Fire precautions Act and the Building Regulations

Background to the forthcoming changes in fire safety legislation

Currently there are two legislative drivers to the fire safety market. New buildings and major refurbishments have to conform to the Building Regulations which require both emergency lighting and fire alarms in most buildings. The building inspectors enforce these regulations during the construction phase and this protection will remain in place.

Introduction

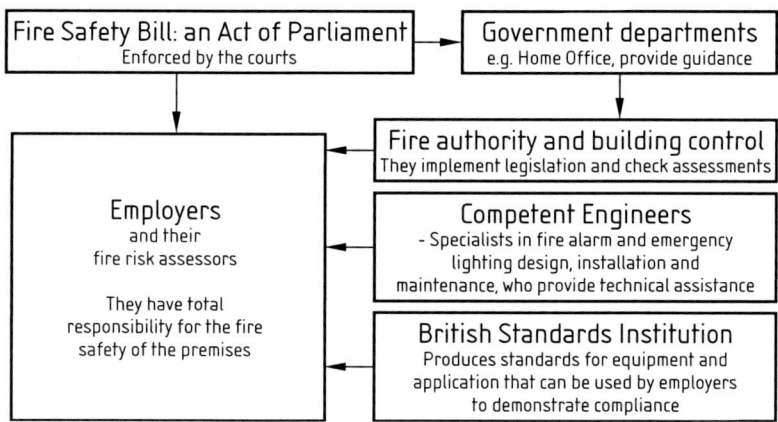

Figure 3 Flowchart of implementation of the Regulatory Reform (Fire Safety) Order and the Building Regulations

The changes defined in a new White Paper will replace all other fire safety legislation with a single act. Unfortunately, this means that the inspection of buildings and issuing of a fire certificate will be limited to a very small number of high risk premises. Other buildings will rely on their employers being required to conduct a risk assessment and make whatever changes are necessary. The advantage of this new procedure is that it will be retrospective and so if a premises had a fire certificate issued 20 years ago it will have to be reassessed and be upgraded to the latest issue of the standards.

The outline of the procedure for employers with sites that are not classified as very high risk is that:

- They must appoint a 'responsible person' who has to provide the risk assessment.
- This person has to evaluate the hazards and the people at risk and then ensure that the fire precautions are adequate for that risk (meeting standards is deemed as complying).
- They are encouraged to use specialist competent people to assist them in providing reports on areas of risk and protection as needed.
- Written records must be kept for all premises with more than five employees.

Implications of the implementation of the Fire Safety Order

With the removal of the enforcement of fire certificates for existing premises, it is important that employers understand their responsibilities to ensure that adequate emergency lighting and fire alarms are installed to safeguard occupants by understanding and conforming to the Fire Safety Order. New buildings will still have the support of building control officers to enforce the building regulations.

Employers will benefit by using the services of trained and competent engineers to assist in the design, installation and maintenance of fire protection equipment to ensure the quality, reliability and long life of the equipment. They should use approved third-party certified safety equipment.

Competent engineers should be able to give advice on how to provide systems compliant with the requirements of the relevant codes of practice for emergency lighting and fire alarm system design.

For normal applications, within the scope of the relevant application standard, the employer should be able to accept compliance with current published standards for system design as a basis on which to incorporate the system design into their risk assessment in order to limit the risks to negligible or tolerable levels as required by the legislation.

Where the construction or operation of the building is such that increased hazards are present, or the risk of harm is higher than for a normal application, the employer (in conjunction with specialist advice where appropriate) must determine what, if any, additional measures are to be taken to limit the risks to negligible or tolerable levels.

Definitions

To be able to interpret standards and guidance documents correctly, it is important to understand the meaning of the major terms used in the industry. Unfortunately, not all the definitions in the standards are consistent or easily understood. So the following list gives the basic meanings of the terms.

NOTE The International System of Units is used throughout this book, for example:

s = seconds
h = hours
lx = lux
A = Amperes.

Introduction

'A' national deviations

Within European standards, countries are allowed 'A' deviations to contract out of sections of the standard if they have national legislation that has different requirements. All of these deviations are listed in that standard for each country, so that any engineers working there know the relevant requirements.

Ballast

Controls the operation of a fluorescent lamp from a specified AC or DC source (typically in the range 2.4–240 V). It can also include elements for starting the lamp, for power factor correction or radio frequency interference suppression.

Ballast lumen factor (BLF)

The ratio of the light output of the lamp when the ballast under test is operated at its design voltage, compared with the light output of the same lamp operated with the appropriate reference ballast supplied as its rated voltage and frequency.

Battery

Secondary cells providing the source of power during mains failure.

Battery capacity

The discharge capability of a battery, being a product of the average current and time, expressed as ampere hours over a stated duration.

Battery sealed (recombination)

A battery that is totally sealed, or constructed so that no provision is made for the replacement of electrolyte.

Battery unsealed (vented)

A battery that requires replacement of the electrolyte at regular intervals.

Central battery system

A system in which the batteries for a number of luminaires are housed in one location, usually for all the emergency luminaires in one lighting subcircuit, sometimes for all emergency luminaires in a complete building.

Combined emergency luminaire (sustained)

Contains two or more lamps at least one of which is energized from the emergency supply and the remainder from the normal supply. The lamp energized from the emergency supply in a combined emergency luminaire is either maintained or non-maintained.

Design voltage

The voltage declared by the manufacturer to which all the ballast characteristics are related.

Deviation

If the designer or installer deviates from the requirements of the code of practice, he or she has to identify the reasons for the deviation and demonstrate that the safety of the building is not impaired in this particular application. All relevant parties must be notified and the details must be recorded in the completion certificate.

Emergency exit

A way out which is intended to be used any time that the premises are occupied.

Introduction

'F' mark

Shows that the luminaire can be mounted on combustible surfaces. It does not show that the luminaire is fire retardant.

Final exit

The terminal point of an escape route, beyond which persons are no longer in danger from fire or any other hazard requiring evacuation of the building.

Illuminance

The luminous flux density at a surface, i.e. the luminous flux incidence per unit area. The unit of illuminance is the lux (lx).

Luminaire

An apparatus, which distributes, filters and transforms the lighting provided by lamps and includes all the items necessary for fixing and protecting these lamps and for connecting them to the supply circuit. It should be noted that internally illuminated signs are a special type of luminaire.

Maintained emergency luminaire

A luminaire containing one or more lamps all of which operate from the normal supply or from the emergency supply at all material times.

Mounting height

The vertical distance between the luminaire and the working plane. It should be noted that the floor is taken to be the working plane for emergency lighting.

A Guide to Emergency Lighting

Material time

This is the relevant time that protection should be available for. It commonly refers to the time that maintained exit signs need to be illuminated for, the material time being when the building is occupied.

Non-maintained emergency luminaire

A luminaire containing one or more lamps, which operate from the emergency supply only upon failure of the normal mains supply.

Normal lighting

All permanently installed artificial lighting operating from the normal electrical supply that, in the absence of adequate daylight, is intended for use during the whole time that the premises are occupied.

Rated duration

The manufacturer's declared duration, specifying the time for which the emergency lighting will provide the rated lumen output after mains failure. This may be for any reasonable period, but is normally one or three hours.

Rated load

The maximum load that may be connected to the system and will be supplied for the rated duration.

Re-charge period

The time necessary for the batteries to regain sufficient capacity to achieve their rated duration.

Self-contained emergency luminaire or single-point luminaire

A luminaire or sign providing maintained or non-maintained emergency lighting in which all the elements such as the battery, the lamp, and the

Introduction

control unit are contained within the housing or within one metre of the housing.

Slave or centrally supplied luminaire

An emergency luminaire without its own batteries designed to work with a central battery system.

850°C glow wire test

Enclosures of emergency luminaires on escape routes must pass this test as specified in BS EN 60598-2-22.

2. Emergency lighting standards: an overview

Structure of standards production and administration

International standards

The International Electrotechnical Commission (IEC) is the oldest of the International Standardization Organizations. It was founded back in 1906 with the support of seven countries. The IEC counts today 48 full members, eight associate members and six pre-associates. It is recognized globally as the provider of standards and related services needed to facilitate international trade in the electrotechnical field. IEC members are not required to adopt the standards issued by this body at national level.

CENELEC members also represent their countries directly in the IEC. They ensure close cooperation and parallel adoption procedures between the technical work going on in the relevant bodies of CENELEC and IEC. The direct source of the reference documents used as a basis for the CENELEC standards is the publication and draft documents of BSI.

The International Organization for Standardization (ISO) was established in 1947. It is a non-governmental, worldwide federation of national standardization bodies with approximately 140 members. Each member represents a single country and is 'most representative of standardization in that country'. The mission of ISO is to promote the development of standardization and related activities in the world, with a view to facilitating the international exchange of goods and services. ISO is very closely related to CEN.

European standards

European standards (EN) are documents that have been ratified by a European standards organization (CEN or CENELEC). A European standard is a document, established by consensus and approved by a recognized body that provides, for common and repeated use, rules, guidelines or characteristics for activities or their results, aimed at the achievement of the optimum degree of order in a given context. Standards should be based on consolidated results of science, technology and experience, and are aimed at the promotion of optimum community benefits. They are designed and created by all interested parties through a transparent, consensual process.

European standards are a key component of the Single European Market. Though rather technical and unknown to the general public and media, they represent one of the most important issues for business. Although often perceived as boring and not particularly relevant to some organizations, managers or users, they are actually crucial in facilitating trade and hence have high visibility among manufacturers inside and outside the European territory. A standard represents a model specification, a technical solution, against which a market can trade. It codifies best practice and is usually state-of-the-art.

The fact that European standards must be transposed into national standards in all member countries guarantees that a manufacturer has easier access to the market of all of these European countries when applying European standards.

British standards

Implementation of an international or European standard that is endorsed and published as a British Standard gives that standard the status of a national standard. Some standards are developed in the UK and are not adopted within Europe, in that case they only have the BS mark of national origin. BSI promote and market the standards within the UK. They also provide representatives to the various European and international committees that draft the standards. When a draft is produced it is sent to each national committee for their comments, which are then integrated into the document by the drafting committee. This final version is then circulated to the national committees for them to vote their acceptance and to identify any 'A' deviations that may be required. When the European standards are available any variations in existing UK standards have to be aligned. For example, the testing of self-contained luminaires was originally once every six months for a third of capacity, but the introduction of BS EN 50172 changed this to an annual test for full capacity, so BS 5266-1 needed to be modified to be consistent with this requirement.

Emergency lighting standards

The increasing complexities of forms of emergency lighting and the need to harmonize standards across Europe have resulted in a wide range of standards covering lighting levels, application requirements and product performance. This chapter is intended to show the outline of the subjects covered by each document and how they interface with each other (see Figure 4).

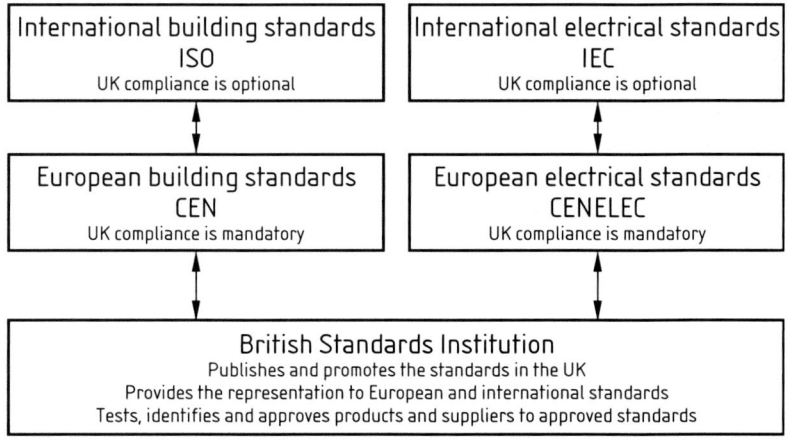

Figure 4 Relationship of standards

BS 5266-1:2005 Code of practice for the emergency lighting of premises

This is the base document that calls up the relevant standards and outlines how they should be interfaced. It also details how types of emergency lighting systems should be designed, installed, commissioned and tested for different types of premises. Installation and wiring considerations are detailed and the appropriate testing and commissioning documentation is defined.

NOTE: This standard is called up as the relevant standard by the Building Regulations, the ODPM and the Health and Safety Executive's publication 'Fire safety: an employer's guide'.

BS EN 1838/BS 5266-7 Lighting applications: emergency lighting

This standard was developed by CEN 169 Working Group 3 based on input from all European countries and is applied to all of them. It supersedes the lighting levels previously given in BS 52266-1:1988.

It details the conditions of illumination needed and the location of luminaires to assist emergency escape from a building. It is split into categories which cover: the escape routes, open areas, high risk task areas and lighting conditions for the visibility of safety signs. Each category has specific minimum levels of horizontal illumination in the worst case during the defined discharge duration period. To ensure that the lighting is satisfactory, limits of uniformity, disability glare and colour are given, in addition maximum viewing distances are provided for the use of illuminated safety signs.

Compliance with this standard is required by BS 5266-1 and can be demonstrated either by measurement, which is laborious, or the design can be done in advance using authenticated data on approved luminaires from schemes such as the Industry Committee for Emergency Lighting (ICEL) registered scheme for BS EN 60598-2-22 luminaires.

BS EN 50172/BS 5266-8 Emergency escape lighting systems

The design section details the consultation procedure, use of plans and records and selection of system types and equipment with the methods of operation. It also covers the design principles to be achieved and states that the performance conditions given in BS EN 1838/BS 5266-7 must be met.

BS EN 60598-2-22 Luminaires for emergency lighting

This standard is the European version with specific national variations from the original parent IEC document. It covers both self-contained and slave luminaires for centrally powered systems. The standard is designed to ensure that the products are safe in use, will perform correctly, and will be compatible with other system components.

Specific sections cover the construction, endurance, operation and output performance tests for emergency lighting use. Many of the tests and requirements are those specified for normal mains luminaires in BS EN 60598-1.

The light output tests measure the light in 5° steps on both the transverse and axial plane together with the tests of total light output at the end of discharge. This forms the base technical information from which the authenticated spacing tables referred to in BS EN 1838 are derived.

BS EN 50171 Centrally powered battery power systems

This standard was developed by CENELEC BTTF 62-8 and covers the central battery units and systems except for the wiring. It also defines the safety, construction and performance of these systems. Where possible similar requirements have been used to those called up for the luminaires to obtain a consistent performance and assist users and manufacturers to become familiar with the categories of construction. A major area of application and output performance requirement of these systems relates to inverters where past problems have highlighted the need for care. The areas covered are:

- The need for the luminaires to be compatible with a correctly rated inverter.
- The inverter must be able to blow all distribution fault protection devices in the emergency condition. During an emergency if a fire burns through the cable in a particular fire compartment causing a short-circuit it is important that the inverter has the capability to isolate that area by blowing the fuse and then continuing to supply the rest of the building.
- The inverter must be capable of starting the emergency load in the emergency condition from the battery supply. In the past some systems relied on starting the load from a by-pass on the mains. These could be unable to re-strike emergency circuits after a voltage depression such as occurs while blowing a distribution fuse.

The safety requirements for batteries are given in BS EN 50272-2:2001. This details both ventilation and electrical safety precautions for all stationary battery installations. There are stringent limits for systems with batteries above 120 V nominal, particularly for installation and maintenance.

Draft standards in preparation

IEC 62034 Automatic testing systems for emergency lighting

With the increasing cost of labour, automatic testing systems for emergency lighting systems are becoming increasingly attractive. The new standard is being written to ensure that tests only occur when it is safe to discharge the system, that the test units cannot interfere with the emergency operation and that fault signals are clear and cannot be cancelled until the system is rectified.

Pr EN 13032-3 (CEN 169 WG 7) Lighting applications: data presentation

This working group is developing harmonized ways of presenting data for Europe. Within the UK this procedure has already been established for emergency lighting use. Luminaire data, when tested by a national test house, can be submitted for registration by ICEL under their scheme for authenticated data which is maintained as part of their ISO monitored systems. This provides users with easy to use spacing tables which have been derived and independently checked from results reported from test houses.

BS 7671 IEE Wiring Regulations/BS 5266-1 Code of practice

These requirements are currently contained in BS 5266-1 which defines the types of cables to be used for centrally powered systems. The general good principles covered in the IEE regulations are also required. Ultimately, it is hoped that the upgrading of the IEE regulations that is currently underway, will cover all wiring aspects.

Other related standards

BS 5266-2 Code of practice for electrical low mounted way guidance systems for emergency use

This covers way guidance systems which may give a useful additional low mounted indication by means of illuminated markers which designate the escape routes.

Emergency lighting standards: an overview

BS 5266-6 Code of practice for non-electrical low mounted way guidance systems for emergency use

These systems meet the same application as the electrically powered units in BS 5266-2 but use photoluminescent sources, so this standard has extra requirements to ensure that the markers have been provided with sufficient light prior to an emergency to enable them to operate in an emergency.

BS 5266-4 Code of practice for design installation maintenance and use of optical fibre systems

BS 5266-5 Specification for component parts of optical fibre systems

These systems relate to the use of optical light guides to transmit the light from the illumination source to remote light output 'tails' that can be used to provide emergency illumination where it is required.

3. Code of practice for the emergency lighting of premises (BS 5266-1:2005)

Introduction

The introduction of BS 5266-1:2005 explains that 'The aim of this standard is to promote wider understanding of the different types of emergency lighting system which may be employed and to give guidance on their correct application to the varied requirements of different categories of premises.'

The document provides coverage for all the emergency lighting standards and explains where they should be used. It gives guidance as to the principles that should be adopted and details of specific requirements. It also establishes the details and controls needed for the use of the UK 'A' deviations from the relevant European standards.

The standard does not itself declare legal requirements. The standard itself is not a legal requirement but it is called up by government guidance documents on how to comply with the legislation. Consequently if systems meet the recommendations of BS 5266-1 they are deemed to conform to the relevant legislation. So, although it is not absolutely mandatory to follow the standard, it does establish a proven and acceptable procedure for providing emergency illumination and safeguards users, designers and installers by enabling them to demonstrate that they have meet their legal obligations.

The use of risk assessment techniques promoted by the Regulatory Reform (Fire Safety) Order establishes the areas that will require application of the standard. Its use encourages uniformity of application, based on providing adequate safety to persons in the event of an interruption of the normal lighting and having due regard to the hazard level and degree of familiarity of the occupants with the particular premises. The standard recognizes that, in addition to ensuring safe, unobstructed means of escape from the premises at all times, an important function of emergency lighting is to make possible the immediate location and operation of fire alarm points and fire fighting equipment. Another function is to minimize the chance of panic arising in large numbers of people or in enclosed spaces, such as lifts. It helps to deter opportunistic thefts and also provides assistance to occupants in the event of any failure of normal lighting supply reducing the risks of injury by falling or stumbling if they are plunged into darkness.

Changes to the standard

Many of the principles established by the UK over the years in BS 5266-1 were presented to the CENELEC standards body for their adoption. The European committee accepted and developed the document and it has now been published as the application standard for emergency lighting (BS EN 50172/BS 5266-8). It was considered that the best way to present information on these changes was that the BS EN standard should be dual numbered to show its relationship to BS 5266-1 and the location of these new requirements was defined by a new edition of BS 5266-1.

The reason for developing this work in Europe was to try and produce an equal and consistent level of safety in all European countries. Because our European counterparts were convinced that it was desirable for all of us to adopt many of the UK's established methods of emergency lighting, the need to change our previously engineered systems was avoided.

The only technical change is the simplification of testing periods, because the standard now clarifies the fact that each compartment should have at least two luminaires. The UK committee strongly supported this change.

Replacement of CP 1007 Code of practice for the emergency lighting of cinemas

This revision enabled other changes to be made. The most important is the adoption of specific requirements for cinemas that allows the new document to supersede CP 1007. The implications of this are that, provided that the cinema has seats that are fixed to the floor, enabling

Code of practice for the emergency lighting of premises (BS 5266-1:2005)

the seat backs to guide people out to the aisles, then the auditorium is not treated as an open area, but the aisles are treated as escape routes. The old CP 1007 standard required that exit signs had two sources of illumination, one from a maintained battery system and one from a mains supply that had to be energized whenever customers were in the cinema. At the time this standard was written this guarded against the sudden failure of tungsten lamps and also the possibility of failure of the rubber insulated wiring. These requirements are no longer needed as modern cables and installation checks reduce wiring problems and fluorescent lamps now give an indication of their impending failure by black ending or by erratic performance.

The major sections of BS 5266-1 now cover the following topics.

Consultation (clause 4.1)

Many inappropriate designs occur because the interested parties did not consult adequately at the start of the process. It is important that all relevant details are available in order to make informed decisions about the system. These include where possible the items listed below.

- Plans of the premises which show the location of the fire safety alarms and extinguishers as well as the normal mains lighting.
- The latest draft of the fire safety risk assessment of the building. This should identify particular hazards and the location of people who are at risk. It should also indicate the standards of maintenance and testing available.
- Details of the national and any local requirements that will have to be met for the particular building application type.
- Information regarding the likely emergency lighting equipment that could be considered for use. This will enable it to be considered for its aesthetic impact on the building and provide its performance to enable the spacing layout to be assessed.
- If the project is an extension or modification of an existing building, details of that installation should be available together with any fire certificate that has been issued and the test records.

Those involved should ideally be: the 'responsible person' for the operation of the building, the fire authority, the building control engineer, the system designer, the system installer and the equipment supplier.

The discussions should identify:

- the function of the building;
- the regulations that it will have to conform to;
- the results of the risk assessment, or if that has not yet been produced, as much information as possible.

The following decisions should then be considered:

- the most appropriate mode of operation of the luminaires and exit signs and the duration for which they will be required;
- the most suitable type of system;
- an action plan to provide a design and installation schedule and to establish the testing and training procedures.

The standard details the following specific information that should be established at the consultation stage to enable an adequate design to be produced.

- The escape routes should be established. They should be adequate for the numbers of occupants but may not include every door out of the building as some may not be suitable.
- The location of the fire call points should be identified. Many will be at the final exit doors so they will be adjacent to a luminaire. But those within the building will need a luminaire located within 2m of them. This highlights their position in an emergency. This also applies to fire fighting appliances such as extinguishers, fire blankets and hose reels.
- The position of fire and other safety signs needs to be established. It also needs to be established whether they are to be back illuminated or to rely on a luminaire which is mounted within 2m and able to illuminate them.
- Any obstructions or hazards on the escape route should be noted. Open areas should be checked for size or any other reason that would require the provision of emergency lighting.
- Each final exit door should be assessed to check if an external emergency luminaire is needed or if the fire authority would accept that street lighting would provide safe illumination.
- Any toilets or changing rooms should be checked. It should be noted if they are: over $8m^2$ in floor area, or any size without borrowed light, or for disabled use.

Code of practice for the emergency lighting of premises (BS 5266-1:2005)

- The position of any lifts, escalators or moving stairways should be noted.
- Motor generator, control and plant rooms should be identified and the extent of emergency lighting to provide safe access to them should be agreed.
- If the building has covered car parks, the walkways and exit routes need adequate illumination.
- If the building has standby lighting, its performance needs to be assessed to ensure that its operation complements the emergency lighting system and does not conflict with it.
- Suitable areas for the location of the central power supply units and routes of low fire risk for the distribution cables need to be established.

NOTE: This subject is covered in Chapter 12 (System design) and Chapter 13 (System selection).

Vision and visibility (clause 6)

We do not observe an object by the light which falls on it but by the light reflected from it to the eyes. Different objects are distinguished by the contrast of the changes in light reflected to the eyes. So, objects which are a different colour to their background are easily visible but if they are of similar colours they are very difficult to distinguish. This situation can be controlled for parts of the fixed installation but it is not controllable with movable items. While the legislation specifies that escape routes are not obstructed, during an emergency the normal discipline may break down and items that were being carried can be left on the route, thus causing a hazard.

The amount of light falling on an object (illuminance) is affected not merely by the power and position of the lamps used for illumination, but also by the reflection from the surroundings. However, during the installed life of the emergency lighting system reflected light cannot be relied on as it would be unreasonable to expect the operator of a building to be limited in their choice of décor colour by the need to maintain emergency lighting levels. Consequently emergency lighting schemes have to be designed for worst case conditions, particularly by excluding the effects of reflectance.

Early issues of BS 5266-1 used the low level of 0.2 lx of emergency lighting on the escape routes as this was the best that was available from many of the self-contained luminaires which were then available and only used a single 2.2 W tungsten lamp giving 22 lm output.

To maximize the effect of this illumination escape routes were supposed to have contrasting colours for prominent edges to vertical surfaces at changes of direction and all potential obstructions or hazards on an escape route in addition to the nosing of the stair treads. As a person's ability to see at low light levels diminishes with age, it was defined that the 0.2 lx value should be increased if the building was used by older people.

During the discussions in Europe regarding the minimum light level that should be used, it was established that the minimum of 1 lx enabled people without visual impairment to see obstructions regardless of colour. It was realized that, while under normal conditions good user practice could keep escape routes clear of obstructions, during an emergency this discipline could not be relied on, so the higher level was adopted.

The UK has a considerable number of buildings engineered to the old 0.2 lx minimum level. Provided that the conditions of contrasting colours was maintained they would be safe, so the UK was granted an 'A' deviation to continue to accept those installations. However, because of the chances of obstructions being left on escape routes during the confusion of an emergency escape, BS 5266-1 recommends that the minimum value of 1 lx on the centre line of the escape route is used, particularly for all new installations.

Minimum illuminance and adaptation

There should be an indication from the risk assessment of the types of people at risk and the nature of the hazard they will face. This will vary from application to application. The age of the occupants is important, not only with regard to the amount of light required to perceive an object clearly, but also the time needed to adapt to changes in the illuminance (visual adaptation). In general, older people need more light to follow an escape route and have longer visual adaptation times.

The maximum period which should be allowed to elapse between the failure of the normal supply and the switch-on of the emergency lighting depends upon the rate at which panic may be expected to mount in a particular building. It also depends upon the time taken to adapt to the new, and normally much lower, illuminance provided by the emergency

Code of practice for the emergency lighting of premises (BS 5266-1:2005)

lighting. The illuminances in this standard have been determined from experience and practical tests.

The actual values for the minimum level of illumination of the escape routes and open areas are now given in BS EN 1838/BS 5266-7. But for clarity they, and the UK deviation, are repeated in BS 5266-1. It also details that emergency lighting should be provided within 5 s of the failure of the normal lighting supply. At the discretion of the enforcing authority this period may be extended to a maximum of 15 s in premises that are likely to be occupied for the most part by persons who are familiar with them and their escape routes.

Additional specific locations requiring emergency lighting

Some of the major specific areas requiring emergency lighting have now been transferred to the European application standard BS EN 50172/ BS 5266-8 but other locations which also require emergency lighting in the UK are still defined in BS 5266-1. These include the areas listed below.

Moving stairways and walkways

These routes should not be used in the event of an emergency but illumination is needed for the protection of occupants who may be on the escalators or other moving stairways when the supply fails. Also, it is human nature that regardless of warning signs, people will still use the route out that they are familiar with, so they still need to be protected.

Illumination is particularly important when the moving stairways are stationary, as the first and last tread heights will not be the normal step height, so they need to be clearly visible to be able to be negotiated safely.

Toilets and changing room facilities

Facilities exceeding 8 m^2 gross area, or any areas without borrowed light and any specifically for disabled use, should be provided with escape lighting as if they were open areas. The standard also contains a note to explain that this requirement does not cover hotel facilities for use by the occupants of a single room even though the combined toilet and bathroom may be without natural light.

Motor generator, control and plant rooms

In the event of a supply failure staff are likely to need to access plant rooms in darkness. These can be dangerous locations, so battery powered emergency lighting is needed so that staff are able move safely. If the only operation that they are likely to need to perform is to operate a switch or check a meter, and there are no dangerous obstructions, the normal open area minimum illuminance of 0.5 lx may be sufficient. If, however, they may need to perform a more complicated task, or if the location has elements of dangerous equipment, the risk assessment may identify the need to use 10% of the normal illumination, as is detailed for high risk applications. These locations also include the winding rooms for the recovery of lifts in the event of a power failure.

Construction of escape lighting luminaires and labelling of categories

This standard and the application standard both require that the luminaires used are of an adequate quality and proven performance. This is demonstrated by conformance to the requirements of BS EN 60598-2-22 (clause 3.4). The categories of operation of emergency lighting systems are given in Annex B of that standard. The detail labelling of categories has been changed to codes because of the increasing need for labels to be understood internationally. The types of luminaires and their test systems have also increased in number. The information is given as codes in a table on the luminaire label. These codes are used internationally and are valid for any luminaires produced to meet the IEC requirements.

The first section of the label uses letters and shows the luminaire type:

> X specifies a self-contained luminaire;
> Y specifies a central supply luminaire.

The second section of the label uses numbers to define the mode of operation of the luminaire. This is obvious for self-contained luminaires. But for centrally powered fittings the code relates to the capabilities of the luminaire, not the way in which it may be used in the systems. Typically they will be able to work in the maintained mode so they will be labelled as 1. Although in practice some may only work on a non-

Code of practice for the emergency lighting of premises (BS 5266-1:2005)

maintained system, this is not a problem. The advantage of the labelling is that it would identify if a fitting only designed for non-maintained use is being incorrectly used in a maintained system. The numbers are given below.

Non-maintained (0)

The lamp only illuminates on failure of the supply, so the emergency circuit must monitor and operate on the failure of the normal lighting in the vicinity. As under normal conditions the lamp is not illuminated, the life expectancy of this component is very good.

Maintained (1)

The same lamp is illuminated both when the supply is healthy and when it has failed. So it does not need to monitor the local normal lighting supply unless the maintained circuit can be switched off. In addition to the locations where this form of lighting is required, it is commonly used to provide both emergency and normal lighting from the same luminaire.

Combined non-maintained (2)

The emergency lamp only illuminates on failure of the supply. Other lamp(s) in the luminaire are only illuminated when the supply is healthy. This mode of operation may be used, where in the normal mains lamps are unsuitable for emergency operation, for example with high-pressure discharge lamps, where a secondary tungsten lamp is used to cover the start-up time of the discharge lamp.

Combined maintained (3)

The emergency lamp is maintained when the supply is healthy and when it has failed. Other lamp(s) in the luminaire are only illuminated when the supply is healthy. This form of operation is common when a multiple fluorescent luminaire is converted for emergency use. The maintained function is normally supplied and controlled by the normal lighting ballast. In the event of a supply failure the emergency module disconnects the normal ballast, and then feeds the lamp from the emergency circuit. To ensure that the emergency circuit works when required it must be fed by the normal local lighting supply. To ensure the safety of maintenance engineers the emergency lamp is identified by a green spot on its lamp holder, as during lamp replacement both emergency and normal supplies may be present. This format of emergency lighting is often used for aesthetic reasons. Care must be taken to design the emergency scheme to allow for the photometric distribution of the host luminaire.

Compound non-maintained (4)

This luminaire, with a facility to run an additional luminaire or projector, only illuminates on failure of the supply. This format is often used with beam projector units, by adding an additional projector to increase the area illuminated. Care must be taken to ensure that the battery capacity is adequate to meet the required duration with the additional load and that the fire protection of the connection to the new projector is adequate.

Compound maintained (5)

This luminaire, with the facility to run an additional luminaire or projector, illuminates the same lamps both when the supply is healthy and when it has failed. Typical instances of use of this system are satellite luminaires used with operating theatre lights.

Satellite (6)

This type of luminaire or projector is designed to be operated from a compound luminaire but has no power supply itself.

The third section of the label details the facilities incorporated in the luminaire as a code letter. None, one, or more categories can be defined.

Test device (A)

This category shows that the luminaire has built-in test facilities, which may be manual or automatic.

Code of practice for the emergency lighting of premises (BS 5266-1:2005)

Remote rest mode circuit (B)

This enables the luminaire to be switched off to conserve battery capacity if it is not required in a supply failure. This disconnection can only be made during the supply failure and there are facilities to reconnect the lights if required. Restoration of the normal power supply resets the control, so the lamps illuminate immediately on the next supply failure. This system is very rarely used in the UK but is common in France and some other countries, particularly if supply failures of longer than 3 h are likely.

Inhibiting mode (C)

This mode enables the luminaire to be prevented from discharging if it is not required. To ensure that this facility does not prevent the operation of the luminaire when it is needed the controls should be interfaced with an essential service. Typical applications are for a substation which is not normally manned. So the inhibiting circuit would prevent discharge of the battery when the building was empty, but it be interfaced with a circuit such as the normal lighting, so that when the building is operated it will be fully operational.

High risk task area luminaire (D)

This is a luminaire that provides full light output within 0.5 s, as is required for this application.

The fourth section of the code gives the duration in minutes as a number. For example: the duration of the emergency mode (in minutes) for a self-contained system is given as:

   10 indicates 10 min duration
   60 indicates 1 h duration
   120 indicates 2 h duration
   180 indicates 3 h duration

NOTE: While systems in the past have been designed for a 3 h or perhaps 1 h duration now there is a need for a variety of durations for high risk task applications which require a duration for as long as this exceptional risk is present. That may be a shorter period of time than the evacuation duration.

An example of the classification codes is:

| X | 0 | AD | 180 |

This would represent the following type of luminaire:

X = self-contained
0 = non-maintained
A = with test facilities
D = suitable for high risk tasks (0.5 s for full light output)
180 = 3 h duration

Additionally, any luminaires used in hazardous areas must be certified to national or international standards by a recognized certification authority, as compliance with temperature limits and explosion protection is likely to be required (see BS EN 60079-14 and BS 6467-2).

Wiring and installation practices

To maintain the integrity of the system, the external wiring and installation also has to be of a high quality. All wiring installations should conform to any statutory requirements and the general rules for good practice contained in BS 7671. In addition, BS 5266 contains detailed specific requirements for emergency lighting systems, particularly for the connections between the central power supply units, the controls and the luminaires on the system.

Wiring of self-contained systems

For self-contained luminaires the connection between the battery and the lamp is contained within the luminaire or adjacent to it (within 1 m) so it is considered as internal wiring and has no other requirements to conform to. However, the installation of the luminaire to the normal supply needs to be considered. If the luminaire is non-maintained it needs to be connected to the same final circuit as the normal lighting in the area in which it is located. This means that, if there is a failure of the final lighting circuit, the emergency luminaires in the area will be activated and provide illumination. If the luminaire is of the maintained type then it will always be illuminated, so there is no need for it to be

Code of practice for the emergency lighting of premises (BS 5266-1:2005)

operated by the local mains final circuit. It can be beneficial if it is fed from a separate circuit, as it will then be able to provide illumination from its second supply, and hence will not be limited in the time for which it can operate.

Wiring of centrally powered systems

The connections to the power supply units from the mains should be as direct as possible to reduce the possibility of failure or other devices interrupting and causing unnecessary loss of supply. This is particularly important for maintained systems.

The wiring between the supply and the luminaires should be protected from the risks of interruption by fire and should either possess inherently high resistance to attack by fire and to physical damage or have other procedures to protect the supply. The protective procedures could include: enclosing the wiring in suitable conduit, trunking or building the cable into the structure of the building so as to obtain the necessary fire protection and mechanical strength. It is important that the rules on segregation are observed and that the wiring of escape lighting installations is exclusive to the installation. This should be separated from the wiring of any other circuits, either by installation in a separate conduit, ducting, or trunking, or if the installations require that common trunking must be used, by separation from the conductors of all other services by a mechanically strong, rigid and continuous partition of fire resistant material.

Joints of cables should be reduced to the smallest possible number and all terminations and other accessories should be such as to minimize the probability of an early failure in the event of a fire. Joints, other than those at luminaires or other system components, should be constructed of materials that will withstand a similar temperature and duration to that specified for the cable. These joints must be appropriately labelled. The wording should be 'emergency lighting', 'escape lighting' or 'standby lighting' as appropriate. To avoid confusion with other services and possible unwanted disconnection for the protection of maintenance staff it should also contain the warning 'may be live'. Each isolator, switch and protective device associated with an emergency lighting system should be situated in a position which is inaccessible to unauthorized persons, or it should be protected against unauthorized operation. Appropriate training should be given to maintenance staff to ensure that they are familiar with its location and operation. Appropriate identification and warning labels should be used. Warning labels should be provided in positions where they can be readily seen and read. The labels should state that switching off the normal supply to an emergency lighting system may not make it safe for maintenance purposes. Such warnings are necessary because, for example, non-illumination of a lamp does not always indicate that a circuit is dead. A circuit that is still live could present a hazard to maintenance personnel. This is particularly relevant in compound luminaires where one lamp is on the emergency lighting supply and the others are on the normal lighting supply.

Test facilities

Each emergency lighting system should have suitable means for simulating failure of the normal supply for test purposes. The testing procedure adopted should be appropriate for the application. While it may be acceptable to switch off the total supply to a village hall at the incoming breaker, this would not be suitable for an old people's home in constant occupation. The facility must be capable of conducting the functional monthly test and the annual full duration test without risk to the operator and without unduly impairing the emergency protection of the building. This means that the duration test should only be undertaken, either when the premises will be unoccupied for the period of the test and for recharge of the batteries or self-contained luminaires can be tested alternately if the building is always occupied. Central battery systems with appropriate industrial batteries can be tested by discharging for two-thirds of the rated duration to a higher than normal discharge voltage.

Manual test facilities should be protected against unauthorized use. They should not disconnect any supplies to other essential services. Test switches must be adequate and safe for the application. Fuses can be used to isolate a circuit once it has been switched off but they should not be used to break a live supply. When testing a central system it is important to ensure that the full load is applied so any hold-off relays must be overridden and their operation checked. If the system is of AC inverter type, and there are dimming circuits on the luminaires, these must be set for full output. The process that sets the dimming to full output should be designed to ensure that this will also happen in an emergency.

Automatic test systems for battery powered systems should conform to IEC 62034 and need to be designed to perform the duration test at periods of minimum risk. The site should be assessed to check if, each year at the preset time, the premises will be empty while the test is performed and the batteries are recharged. If this is the case then the accurate timing and protection of the timing circuit against supply failures of up to seven days required by the standard will perform the test at safe times. If, however, the building is always likely to be occupied, self-contained systems need to be arranged so that the timing of the tests of alternate luminaires in the building are offset by at least 24 h. This provides worst case illumination from at least one luminaire in each compartment of the building if the emergency occurs during the timing cycle. Central systems normally use a single battery so this procedure cannot be used but fortunately the characteristic of the stationary batteries used enables them to be partially discharged for two-thirds of their capacity. Then, provided that the output voltage is above a preset value which is higher than for normal discharges, it will confirm that the battery is in good condition while still retaining enough capacity to evacuate the building if an emergency were to occur during the test.

Voltage compatibility of a slave luminaire and a central battery system

The following considerations should be taken into account to ensure that a slave luminaire is designed to be able to work over the variation of output voltage range from the central power supply to which it is connected.

- For DC systems, the battery output will vary from the initial fully charged battery voltage down to the end of duration voltage when the battery has discharged at the end of the duration.
- AC supplied from the mains may possibly have to cope with the new wider tolerances of 230 V ±10%.
- For AC supplied from a step down transformer, in addition to the supply variation the load regulation of the transformer needs to be included if varying partial loads are ever likely to be encountered, as for instance if the system has a number of hold-off relays on the maintained circuit. For sensitive applications, such as hospital theatre lights, tapped outputs can be used.
- For AC supplied from an inverter, provided that it complies with BS EN 50171, the output will be regulated to ±6% for a constant load with excursions of up to 10% for load variations from 20% to 100%.

All systems also need to accommodate cable voltage drops which are allowed to vary from almost nothing at the supply unit to a reduction of 4% of the system voltage at the furthest luminaire.

The most difficult combination is for maintained luminaires when they may have to accommodate two different supply tolerances plus the cable voltage drop. The system designer has to ensure that the power supply output is within the luminaire voltage capabilities.

Photometric data should be derived from the lowest voltage that can be applied to the luminaire.

While the minimum time for evacuation is defined in BS EN 1838:1999/BS 5266-7:1999, as being 1 h this is only acceptable for the few premises which will be emptied immediately on a supply failure. These systems should not be re-occupied until the battery has recharged, which normally takes 24 h. Consequently most systems in the UK provide a 3 h duration which allows most supply failures to be bridged. On the few occasions that the supply outage is longer, this enables operators to ensure that the building is safe either by evacuating people after 2 h or, as in the case of a hotel, that the guests are safe in their rooms.

Guidance on the category of system to be adopted for typical premises

The standard gives specific indications for the most appropriate category of system. This should be reviewed together with the information on risks and hazards for individual premises revealed by the risk assessment. For many types of premises there are statutory requirements relating to emergency lighting. It is important that the appropriate authority is consulted.

Premises used as sleeping accommodation (clause 10.3.2)

This type of application is known to be a high risk. It includes premises such as: hospitals, nursing homes, hotels, guest houses, clubs, colleges and schools. Many of the occupants are likely to be unfamiliar with the escape routes so adequate illumination with clear and unambiguous signage has to be readily visible in an emergency. Because it is unreasonable for the operator to evacuate the building in a supply failure, a 3 h system would normally be used (see clause 10.4a).

Code of practice for the emergency lighting of premises (BS 5266-1:2005)

Non-residential premises used for treatment or care (clause 10.3.3)

This type of premises is playing an increasingly important role and the medical procedures being adopted are becoming increasingly sophisticated and lengthy. The time to terminate a treatment and to prepare a patient to leave is also likely to mean that most premises will benefit from a 3 h duration system. In treatment areas the risk assessment will indicate the level of illumination needed and the changeover time in break of illumination that would be acceptable for each area.

Non-residential premises used for recreation (clause 10.3.4)

These premises tend to accommodate large numbers of people who may be unfamiliar with the building. Different applications have their own problems. These include: sports centres where both the participants in the sport and the spectators need to be protected. For sports stadia reference should be made to the legislative guidance for these premises.
Exhibition halls are likely to have varied layouts of escape routes and obstructions, also the entire floor areas could be occupied. So the provision of emergency lighting should enable exits to be visible and accessible from any part of the floor. If individual exhibits at a particular show obscure the illumination of the routes or signs, temporary arrangements may have to be made.

Premises such as restaurants and public houses which provide food and drink were regarded as a special risk because of the influence of alcohol and the fact that to generate atmosphere the normal lighting is often of a low level. Originally, maintained emergency lighting was called for to ensure that minimum safe levels were provided. But it is now understood that the operational requirements will keep a safe level. Consequently, maintained operation is limited to the signs always being illuminated, so that they are conspicuous when the building is occupied.

Entertainment centres such as theatres, cinemas and concert halls often require very low levels of illumination in the auditorium in their normal operating conditions. These levels are then increased in the interval or for end of the performance. To ensure that safe levels are provided, maintained operational lights are used to cover the area to 0.1 lx at a plane 1 m above the floor (equivalent to the height of the seat backs). Provided that the cinema or theatre seating is fixed to the floor, the seating rows will channel people to the aisles, so these auditoria do not need to be treated as open areas. For these applications, provided that a very low level of overspill light reaches the backs of the chairs, people will be able to work their way along the row and then use the aisles as the normal escape route.

Normal lighting levels for escape routes should be provided in the aisles by additional lighting in the event of a supply failure and the need to evacuate.

Non-residential premises used for teaching, training and research, and offices (clause 10.3.5)

These premises, such as schools, colleges, technical institutes and laboratories, tend to be mainly low risk. However, some areas may contain risks which need extra protection. For example, while most classrooms are smaller than 60 m2 and so do not need illumination, if they are used as a chemistry laboratory, open area emergency lighting could be needed.

Non-residential public premises (clause 10.3.6)

This type includes premises such as: town halls, libraries, shops, shopping malls, art galleries and museums. Municipal premises may not require immediate re-occupation and often use central systems which can readily be supplied as 1 h duration. But the commercial pressures on shopping malls would be likely to need a 3 h system so that trading could continue as soon as possible.

Industrial premises (clause 10.3.7)

Industrial premises are those used for the manufacture, processing or storage of products. These factories, workshops, warehouses and similar establishments often house high risk task areas. Their control rooms need 10% of the normal mains illumination provided within 0.5 s. In some plants generators are used to provide standby lighting to enable production to continue through power failures. But it is not normally economic to upgrade them to provide emergency lighting. The generators must be able to provide illumination within 5 s (or the fire authority may allow that to be extended to 15 s if the occupants are familiar with the building). However, they will still need to be wired in fire resistant cable and final circuit failure protection must be provided. So it is preferable to use the generator for standby duty and add battery fed emergency light for 1 h duration to protect the premises until the generator has powered the normal lights. In this case the generator should then also supply the emergency lights, enabling them to recharge ready for future operation. Emergency lighting units can be fitted with a timer to keep their lamps illuminated until the high pressure discharge lamps have re-struck after restoration of the normal supply.

Multiple use of premises (clause 10.3.8)

Difficulties occur when a premise has a number of different applications. The standard recommends that they should all meet the level of protection of the most arduous application. For example, if a shopping mall has a restaurant within it, then the 3 h duration required for the restaurant should apply to the whole building.

Common access routes within multi-storey dwellings need to be provided with emergency lighting although the individual apartments do not need it.

If access for the fire brigade is difficult, for example, if the premises are above the reach of fire platform appliances or are in a basement, the 3 h duration should be used. In this case, the function of the emergency lighting is to reassure any occupants who may be trapped in the event of a fire in another part of the building.

The walkway areas of covered car parks should be provided with escape route emergency lighting. The luminaires selected should have an adequate IP rating for weather protection, they may also need to be of vandal resistant and of appropriately rugged construction.

Design of system

When the details of the project have been established the design of the emergency lighting installation can proceed. Guidance is given as to the steps that should be followed to complete the design. The aspects that should be covered are as follows.

- Decide the type of system of emergency lighting to be used.
- Establish whether the luminaires are to operate only in the event of a supply failure (non-maintained), or if they are also to be illuminated when the supply is healthy, and if so what level of illumination is needed for the normal lighting (maintained or combined).

- If a central powered system is selected the wiring needs to be considered. This includes: selecting the type of cable and its route, the safety provisions and the cable voltage drop calculations. The compatibility between the power supply and the luminaires should also be confirmed.

Design of illuminance

Having determined the positions and areas which need to be illuminated for the emergency lighting system, the detailed design can commence.

The layout of escape routes and location of open areas and other areas that need protection should be marked, and the mounted height of the luminaires and their mode of operation should be determined.

Appropriate luminaires should be selected and the manufacturer's photometric 'authenticated data' obtained. This may be in the form of a spacing table or as basic data for a computer program or manual calculation. Most schemes are calculated with a 20% allowance for the effect of dirt on the luminaire diffuser. This should be checked to see it is suitable for the particular application.

The emergency luminaires should be located on the escape routes shown on the plan at the points of emphasis given in BS EN 1838/ BS 5255-7.

The routes and open areas should then be checked to see if the minimum illumination required by the standard has been achieved. If not, then more luminaires should be added until their spacing meets the required limits.

The design of the testing procedures, circuit protection and controls requires consultation with those responsible for the continued operation and maintenance of the system.

When the design is complete fully detailed plans, with the location and details of the emergency lighting equipment, should be produced. They should be followed by the installer and those performing the verification of the system.

The designer should include the preparation of instructions on the operation, testing and maintenance of the system in the design schedule.

Completion certificate

Particularly with the decreasing use of fire certificates, employers will now need to be able to demonstrate, by a completion certificate, that their installation was correctly designed and installed.

Code of practice for the emergency lighting of premises (BS 5266-1:2005)

The certificate is now split into four sections so that engineers only endorse those areas of work that they have been responsible for. The first section is the overall declaration that each section has been correctly completed and draws together any deviations from the standard.

The second section covers the design that has been produced and gives evidence that adequate photometric calculations from authentic data are available; this removes the need to conduct difficult and time consuming tests on site.

The third section enables the installer to confirm that the design that was produced was correctly installed.

The final section is to be completed by the engineer who conducts the verification tests that the system works as designed.

Deviations

BS 5266-2:2005 is a guide on good practice for typical applications, so an individual site may have a valid case for deviating from the standard. But, if this is done, it should be identified so that all who are involved with the project are aware of the reasons for it. These deviations would normally need justifying by information from the risk assessment which endorsed that safety would not be impaired. A typical example may be that for a ground floor corridor in an office block, defined by the risk assessment as low risk, a luminaire spacing of 10% greater than the authenticated data may be regarded as acceptable. If so, the deviation should be noted in the completion certificate. Then at any future time it will be a justification and record of why the deviation was accepted.

Servicing should then be performed in accordance with the manufacturer's recommendations. This is particularly important for generators and centrally powered systems.

Routine inspections and tests

Details of the test regime are detailed in BS EN 50172/BS 5266-8. BS 5266-1 contains a typical test log which enables the operator of the building to not only record the test results but also to monitor the action to have the system repaired. The log also gives guidance on the action that should be taken to reduce the risks to the premises while the emergency lighting is being repaired. To ensure that intervals without a fully effective system are minimized, it is recommended that comprehensive spares are kept on site and that the operator knows how to contact competent service organizations.

Where the premise contains a large numbers of occupants, or where adequate local maintenance staff may not be available, consideration should be given to the use of an automatic test system. An automatic test system for battery powered systems is specified in IEC 62034.

The appendix of BS 5266-1:2005 contains a template for typical completion certificates and test log records. They are also shown in Annexes A and B of this book.

4. Emergency lighting (BS 5266-7/ BS EN 1838)

Introduction

Escape route lighting is designed to provide appropriate visual conditions and direction finding to enable occupants to find and use the escape routes to a place of safety. It also ensures that fire fighting and safety equipment can be located and used.

The open area illumination is provided to enable people to leave rooms and reach an escape route. Its presence can significantly reduce the chances of panic occurring in an emergency.

High risk task area lighting is used to enable people to shut down a dangerous process in the event of an emergency, or to avoid dangerous equipment particularly at the instant of loss of supply. For this reason it provides a proportion of the normal illumination and not just a minimum safe value.

For the escape route and the open area lighting, the working plane for emergency lighting is considered to be the floor. But for high risk task areas it is the area of the hazard, for example the cutters of an unguarded milling machine. The requirement for high risk task lighting would normally be identified by the user's health and safety risk assessment.

General principles

Emergency lighting not only provides illumination but the luminaires also act as beacons that are used to emphasize the location of hazards and safety equipment. This technique may be applied to cover particular problem areas in specific applications, but all premises should cover the locations that are listed in the standard. These are given below.

- At each exit door that is intended to be used in an emergency. This means that doors which are not intended to be used would not be illuminated so as not to encourage their use, perhaps because the route that they would access might be unsuitable.
- Near to stairs. To enable these to be used safely, it is important that occupants should know the location of each tread. So the luminaires must all direct the illumination. A fitting is also required near to any other changes of level, such as steep ramps or a single step.
- All exit signs must be illuminated. Any other safety signs that identify a hazard that would still be a significant hazard in a supply failure, for example the presence of a radioactive source, must also be illuminated.
- Illumination must be provided at each directional change in the escape route. As well as indicating the extent of the route this requirement promotes good engineering as it allows the light from the luminaire to shine down both corridors.
- The same applies to the intersection of corridors on the escape route and has the added benefit of providing the maximum illumination at the point where two streams of people exiting the building converge. This will help to reduce confusion in these potentially congested areas.
- External illumination near to the final exit is needed to safeguard escape to a place of safety as people are still at risk until they are away from the influence of the building.
- A new requirement to the UK was the requirement to illuminate each first aid point. This is important as, in the event of a supply failure alone, if a person injures themselves their priority is to obtain treatment not to leave the building. If the first aid area is not adjacent to the escape route or part of a protected open area the floor should be illuminated to 5 lx.
- Emergency luminaires are also required near to each piece of fire fighting equipment and fire alarm call points. These locations also need an illumination of 5 lx if they are not adjacent to the escape route or are not part of a protected open area.

NOTE: for the above requirements the term near is defined as: needing an emergency luminaire placed on either ceiling or wall and being within 2 m measured horizontally.

Emergency lighting (BS 5266-7/BS EN 1838)

These requirements apply to the whole building. If compliance requires emergency lighting in rooms smaller than 60 m² in area, consideration should be given as to whether or not the risks would justify the whole room being covered by the open area requirements. An example would be a small kitchen which, because hot food was being prepared, would contain an extinguisher and fire blanket. These would need emergency lighting near to them.

Light levels must be designed with adequate de-rating factors to ensure that the specified values are available over the full duration, taking into account the effects of dirt and aging components at the end of the system's design life.

To avoid the difficulties of conducting photometric tests onsite it is preferable to use tested luminaires. These are available with authenticated data either in the form of spacing tables or as basic data for computer programs or manual calculation.

Escape routes

The escape routes need to be designed by placing the luminaires to cover points of emphasis to cover the hazard points and safety equipment. Then any extra luminaires that are needed must be added to provide a safe minimum level of illumination over the whole route. The minimum illumination required is 1 lx on the centre line of the exit route. At least 0.5 lx must also be available over 50% of the width of the route for routes up to 2 m wide. If the route is wider than 2 m and it is all needed to allow passage of the people evacuating the building, then the route should be treated as a number of 2 m wide strips. If a route of 2 m wide would be sufficient to allow the maximum number of people to evacuate the building, then a 2 m wide escape route could be designated through the area, unless the area was over 60 m² in which case it should be treated as an open area (see Figure 5).

While providing minimum light levels, the illumination along the escape routes should be reasonably uniform and the change from brightly lit to dimly lit areas should not exceed 40:1.

Ratio of maximum to minimum illuminance

This is equivalent to moving from bright sunlight into deep shade and the higher elements of illumination should not affect the eye's ability to observe details at the lower lighting levels.

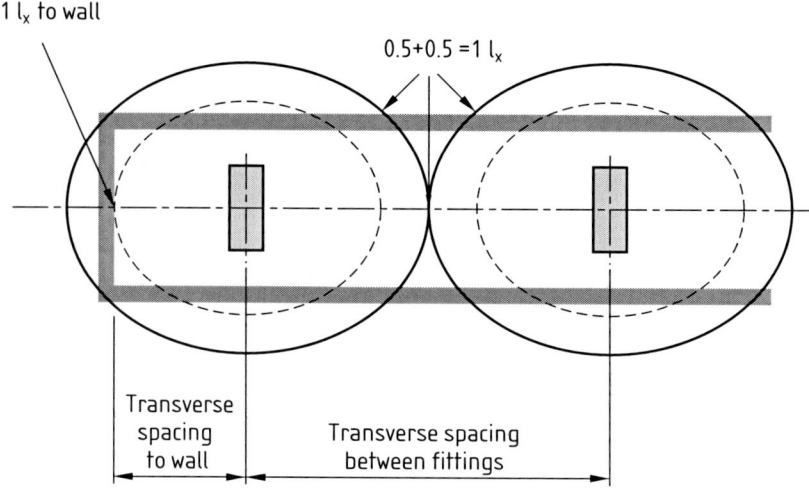

NOTE Illumination of escape routes (Isolux plot) is specified in 6.6 of BS 5266-1:2005 which refers to BS 5266-7 and 4.2.1 of EN 1838:1999.

Figure 5 Illumination of escape routes and isolux plot

Disabling glare from the luminaires in the direct line of sight also needs to be controlled. This is particularly true for installations of beam projectors or high pressure discharge lamps. The standard defines the maximum values that can be accommodated by the eye. These values are identified within a table of specific values for different mounting heights. The table defines the maximum values that can be allowed within 30° of the line of sight when walking along a level floor. If the route has varying height levels such as on a staircase then the values must apply at all angles of view. Even at the lowest value for ceilings up to 2.5 m high the maximum luminous intensity allowed is 500 cd. This allows the use of fully powered fluorescent lights up to 58 W in size.

Because safety signs and extinguishers are colour coded it is important that under emergency conditions these colours are still recognizable. This is achieved by checking that the emergency luminaires provide a reasonably white light. The precise limit for the colour of the lamps is RA 40 or higher. This value is stated as part of the lamp's information. In practice it only normally excludes the use of yellow low pressure sodium lamps. Fortunately, it is also part of the luminaire test, so fittings conforming to BS EN 60598-2-22 will be acceptable.

In the event of a supply failure a fast provision of the emergency output assists occupants to avoid injury and also provides a considerable degree of reassurance in the emergency. Unfortunately some of the

Emergency lighting (BS 5266-7/BS EN 1838)

most efficient forms of light source need time to achieve their full output. Consequently the standard requires that 50% of the minimum emergency level is available in less than 5 s and the full output has to be available in 60 s and maintained to the end of the rated discharge. This condition for the 1 lx minimum on escape routes is different from the old values still covered by the UK 'A' deviation for 0.2 lx which requires that the full output is available within 5 s.

Compliance with these requirements

This is achieved either by measurement or by the use of authenticated data from a supplier. The annex of BS EN 1838 details the type of meter that must be used. The meter must be weighted for the frequency response of the human eye and must be cosine corrected so that it converts light falling on the sensor to the vertical component. This is important as it is common in some countries outside Europe to point a non-corrected meter at the light source, thus obtaining a higher reading than is acceptable to the standard, which calls for the lighting measurement to be that component that falls vertically on the working plane (floor). Measurements also have to be taken without any additional light spillage from any other source. This means that the test needs to take place at night with every window blacked out and it needs to be taken at the end of the discharge period. This is difficult to perform and takes a considerable time. Worse still, if the result is a failure the system would need to be upgraded to conform to the standard. These factors make this procedure unpopular and so it is only rarely used, normally in the event of a dispute as to the system's acceptability. Instead, manufacturers can supply products with third-party authenticated data. ICEL, their trade association, lists luminaires with approved spacing tables and basic data for use in computer programs.

The procedure requires that the luminaires are approved to BS EN 60598-2-22 and that the photometric output and distribution is used to produce and check the spacing tables by proven calculation using the point source methods (see Figure 6).

Open areas

Previous issues of the BSI code of practice used a simple procedure of requiring open areas to be illuminated to a minimum of 1 lx average.

A Guide to Emergency Lighting

This worked reasonably well for luminaires which had an all-round distribution but unfortunately mains luminaires with sophisticated controlled distribution became popular as emergency luminaires. These included tungsten–halogen down lighters with restricted beam angles and fluorescent luminaires with a total cut-off of lighting distribution at 65°. These were designed for use in offices with visual display units but, for aesthetic reasons, they have been used in many other applications. The result of using either of these as emergency luminaires to provide 1 lx average illumination was that the floor would have pools of light under the fitting, then considerable distances of near darkness to the next fitting.

Ceiling mounting height (m)	Escape Routes 1 lx min. along centre line				Open (anti-panic) areas 0.5 lx min. Luminaires in a regular array			
	Transverse to wall	Transverse spacing	Axial spacing	Axial to wall	Transverse to wall	Transverse spacing	Axial spacing	Axial to wall
2.5	1.8	5.6	4.7	1.5	2.0	5.4	4.4	1.6
3	1.4	5.5	4.6	1.1	1.9	5.5	4.6	1.6
4	-	-	-	-	1.6	5.5	4.6	1.3

NOTE Authenticated spacing tables determine whether additional fittings are necessary to those required for the points of emphasis.

Figure 6 Authenticated spacing table for lighting level, showing maximum distance between luminaires

To overcome this unsatisfactory situation the new standard defines minimum levels of illumination of 0.5 lx anywhere in the central core area of the room and also the same 40:1 ratio of maximum to minimum illuminance as for escape routes (see Figure 7).

The central core excludes a border which is 0.5 m from the walls and also the floor under any movable furniture. However, fixed partitions must be accounted for in the design although they are still allowed a border of 0.5 m round them.

Disability glare limits, colour output of the light sources and the response times are the same as for escape routes and so are the compliance procedures.

Emergency lighting (BS 5266-7/BS EN 1838)

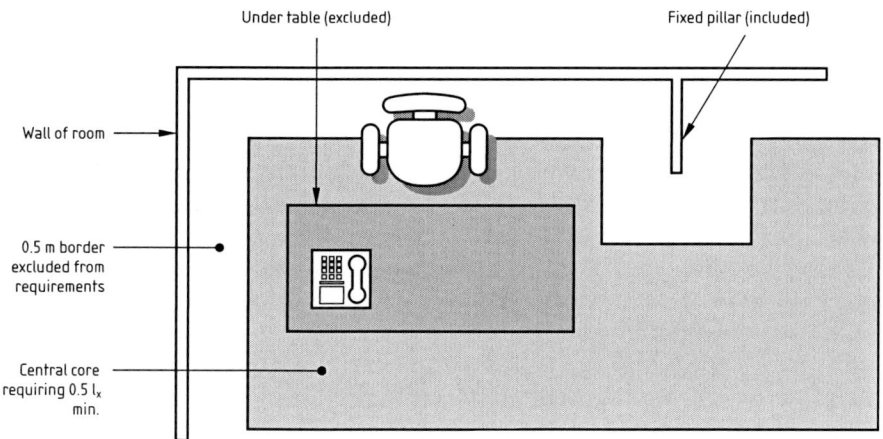

NOTE Illumination of open areas is specified in 6.6 of BS 5266-1:2005 which refers to BS 5266-7 and 4.3 of EN 1838:1999.

Figure 7 Illumination of open areas

High risk task area lighting

An area of concern identified by European legislation is to protect workers from injury if they are conducting a potentially dangerous operation when the normal lighting fails. The lighting has to not only protect workers but also be used in those areas where dangerous processes are controlled. This condition is considerably more arduous than protecting a guest walking along a hotel corridor, so different types of lighting are needed. The area of risk may not be stumbling on the floor but could be at workbench or any other height. Provided that the normal lighting has been checked and proven safe for continuous working, a reduction to 10% of that level will enable the operator to safely terminate the activity.

The emergency lighting requirement is given as a percentage of the normal lighting level, which is defined as an average. Consequently, the emergency value required is also an average and so it is possible to use the normal format of uniformity. The uniformity of emergency illuminance required is that the average value to minimum shall not be less than 0.1.

Maximum values of disability glare reflect the higher levels of illuminance that these systems operate with, so higher values are given for this application, but the principles of their use remain the same.

The duration that these luminaires will be required to provide varies according to the application. The requirement is that the illumination should be provided for as long as the risk is present. If the hazard is a large rotating lathe, while it will be dangerous at the instant the supply fails, after a period of perhaps five minutes the work piece will have stopped rotating and it becomes no more dangerous than any other inert object in the work place. However, if the hazard is a large acid bath that must be accessible as part of the manufacturing process the hazard will remain the same throughout the evacuation period, so this is the duration need from the high risk task luminaires.

Another difference for this type of lighting is that it should provide continuity of vision, so the full design emergency illuminance must be available within 0.5 s. The requirements for lamp colour and also the procedures for demonstrating compliance are the same as for the escape route and open area lighting, but the luminaires must have been tested for light output after 0.5 s. Because not all emergency luminaires are suitable for this duty the manufacturer of suitable fittings would identify to the test house that they wanted those values confirmed. In practice this duty is performed either by high power tungsten beam units or maintained fluorescents as it is unlikely that a non-maintained fluorescent lamp can meet the 0.5 s output limit.

Safety signs

It is important that safety signs are visible at all times that the premises are occupied. In the emergency condition they need to give a clearly recognizable indication of the escape routes, first aid points and any other appropriate safety signs that may have been identified. Because some signs that could be identified as needing illumination in a supply failure are additional to the fire exit and first aid signs they are of different colours so these are referred to in the standard as the contrast colour. The colour has to conform to limits given in ISO 3864. For exit signs this means they must be white pictograms with green as the contrasting colour. The luminance of any part of the green contrasting colour must be at least 2 cd m^{-2} in the worst case and the ratio of maximum to minimum luminance of either white or green shall be not greater than 10:1 to prevent areas of high luminance dominating the sign and impairing the recognition of its message.

Emergency lighting (BS 5266-7/BS EN 1838)

Also, to ensure that the message is clear, there are contrast limits for the luminance of not less than 5:1 and not greater than 10:1 between any junctions of the white and green colours (see Figure 8).

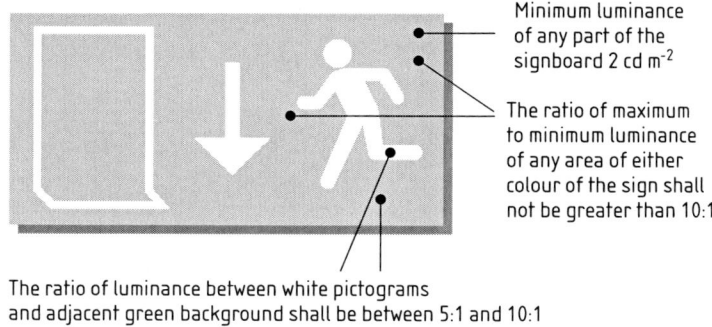

Minimum luminance of any part of the signboard 2 cd m^{-2}

The ratio of maximum to minimum luminance of any area of either colour of the sign shall not be greater than 10:1

The ratio of luminance between white pictograms and adjacent green background shall be between 5:1 and 10:1

NOTE Requirements for luminance of safety signs is specified in 6.6 of BS 5266-1:2005 which refers to BS 5266-7 and 5.1 to 5.5 of EN 1838:1999.

Figure 8 Luminance requirements for safety signs

Maximum viewing distances

To make sure that the sign is readable, limits have been placed on the maximum viewing distance over which each size of sign can be seen in emergency conditions. Although either type of sign can be used, back or edge illuminated signs are conspicuous and usable over twice the distance of the same sign that is illuminated from a remote emergency light which must be within 2 m and able to meet the luminance requirements of the sign. The distances are given as a factor times the height of the green panel and not the symbols. For back illuminated signs the factor is 200 times the height of the sign and for remotely illuminated signs it is 100 (see Figure 9).

Compliance

For those exit signs manufactured with their own source of illumination, either shining through a panel or entering the panel at the edge and illuminating the legend, the sign can be tested against the requirements of BS EN 60598-2-22. Checking the performance of a remotely illuminated sign is much more difficult but it is important that the luminaire should be in the direct line of output to the sign as well as being within 2 m of it. Annex A of BS EN 1838 defines the detail of the testing procedures for signs, including that the measurements should be taken with a 10 mm diameter patch at positions up to 15 mm from the junction of the white and green colour.

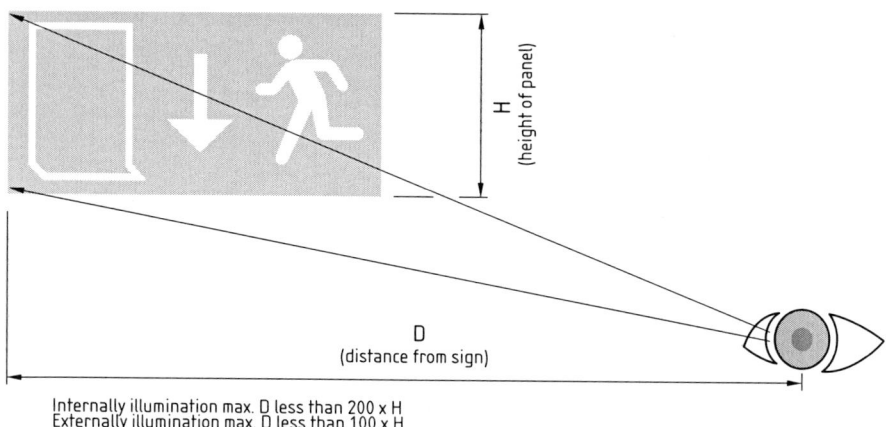

Internally illumination max. D less than 200 x H
Externally illumination max. D less than 100 x H

NOTE Viewing distances for emergency signs are specified in BS 5266-1 which refers to BS 5266-7 and 5.6 of EN 1838:1999.

Figure 9 Viewing distance for emergency signs

'A' deviations on BS EN 1838 applicable to the UK

The UK has three listed 'A' deviations. The most important is that, under certain specific visual conditions, for routes that are to be permanently unobstructed, the old lower level of 0.2 lx is still allowed. This prevents systems that were previously installed being regarded as non-compliant and potentially needing to be modified to meet the Workplace Directive. The UK committee, however, recommends that the 1 lx value should be used as it provides visual conditions under which obstructions that may have been left on an escape route can be seen in an emergency. This is also very useful in applications where the building is always occupied and testing has to be performed by switching alternate luminaires. If an

Emergency lighting (BS 5266-7/BS EN 1838)

emergency occurs during this period with only half the luminaires fully operational the higher level becomes particularly valuable.

Secondly, to retain the acceptability of some systems that are already in place, the UK allows a relaxation of the response time from 5 s to 15 s. But this is only for those premises where the occupants are familiar with the building and the fire authority has given its express permission.

The final deviation related to justifying the use of CP 1007. This is no longer relevant as CP1007 has been cancelled as it has been absorbed into BS 5266-1:2005.

5. Application standard (BS EN 50172/ BS 5266-8)

This standard endorses many of the items which have traditionally been governed in the UK by BS 5266-1, so that code of practice was revised to align with the European norm. Because BS EN 50172 is a standard, not a code of practice, it is normative and so requirements are called up by the term 'shall'. The code of practice was only advisory so it used the term 'should' although in operation it was often treated as though it was mandatory because of the need to maintain safety.

This European standard specifies the provision of illumination of escape routes and safety signs in the event of failure of the normal supply, and specifies the minimum provision of such emergency lighting based on the size, type and usage of the premises. This standard relates to the provision of electric emergency escape lighting in all work places and premises open to the public. It does not cover private domestic premises, but its provisions may be applicable to common access routes within multi-storey dwellings.

General guidance (clause 4.1)

The standard endorses the requirements that have been used in the UK, that emergency lighting must operate in the event of a final circuit failure of the local normal mains lighting (greater detail is given in clause 5.2). It also requires that designs meet the full details of levels and measurement of illuminance and adaptation that are specified in BS EN 1838 and that the installation must conform to the wiring rules of HD384 (in the UK this is often known as the IEE Wiring Regulations).

Identification and illumination of emergency exit signs (clause 4.2)

The standard provides useful guidance on the location of suitable signage, it defines that: where direct sight of an exit is not possible and doubt may exist as to its location an additional directional sign shall be provided. In addition an exit or directional sign shall be in view at all points along the escape route.

All signs marking exits, or escape routes in particular premises shall be uniform in colour and format, and the illumination shall conform to BS EN 1838.

Maintained exit signs should be considered for applications where occupants may be unfamiliar with the building as they are more conspicuous when the normal supply is healthy. Exits illuminated at all times assist people new to the building to recognize the nearest exit to them if they need to evacuate the building. This should be checked with the risk assessment of the building to understand the types of visitors to the building and the ease of their evacuation.

Open area (antipanic) (clause 4.4) (also covered in clause 3.4)

The size of room that requires emergency lighting is now defined as being larger than 60 m^2 floor area or smaller areas if there is an additional hazard such as use by a large number of people.

This clarifies the areas that need protection to BS EN 1838/BS 5266-7. Previously the code of practice just referred to large areas, but this was obviously imprecise. Instead the finite value gives a clear indication of whether the room needs emergency lighting or not. This requirement aligns with the latest issue of the Building Regulations and so brings all buildings into line with the guidance for new construction.

Factors that would require smaller areas to be illuminated include:

- To provide illumination for open areas which have an exit route passing through them to allow safe movement towards and through the exits to a place of safety.
- To ensure that fire alarm call points and fire equipment provided can be readily located and used.
- To permit operations concerned with safety measures.

Application standard (BS EN 50172/BS 5266-8)

- To protect occupants from any fire safety hazards that may have been identified by the risk assessment of the building, such as it being used by a large number of people.

System design (clause 5)

Plan of the premises (clause 5.1)

The need for detailed information on the building is defined, but this is also considered in more detail in BS 5266-1.

Failure of supply to part of a premises (clause 5.2)

Emergency lighting circuits have to operate in the event of local lighting supply failure and arrangements have to be made to ensure that this happens. Self-contained and combined non-maintained luminaires just need to be connected to the local lighting final circuit so that an interruption of the circuit protective device will automatically activate the luminaires. Central battery units and generators will need monitoring and the use of control relays if they are operating as non-maintained systems. In many cases it is preferable to use a maintained system held off by a relay controlled by the final lighting circuit.

System integrity (clause 5.3)

Provision of highly reliable emergency escape lighting is essential. The illumination by the emergency escape lighting system of a compartment of the escape route shall be from two or more luminaires so that the failure of one luminaire does not plunge the route into total darkness or make the directional finding effect of the system ineffective. For the same reason two or more luminaires shall be used in each open area. This requirement clarifies the guidance originally given in BS 5266-1 and also extends the principle to cover open areas. As all compartments now have at least two luminaires it is feasible to test installations that are still occupied by testing and then recharging alternate luminaires before testing the remainder (see Figure 10).

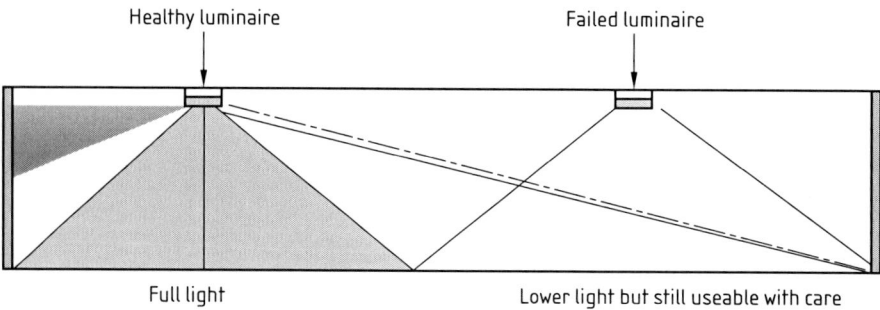

NOTE System integrity is specified in 7.4 of BS 5266-1:2005 which expands the requirements of BS EN 50172:2004 and 5.3 of BS 5266-8:2004.

Figure 10 System integrity

Luminaire quality (clause 5.5.1)

The standard requires that luminaires shall be designed and constructed in accordance with BS EN 60598-2-22 and that they shall have housings that are appropriately protected for their location.

The product standard defines the requirements for construction of housing, charger, inverter lamp and most importantly battery life and the claimed performance data are also verified. The product standard also checks the photometric performance. Without this information difficult field tests would have to be made to ensure that the photometric requirements are met.

Central power supply systems (clause 5.5.2)

Battery powered central power supply systems must be designed and constructed in accordance with BS EN 50171 and the battery safety requirements of BS EN 50272-2.

As most central battery systems rely on a single power supply unit in each area the reliability of the system and the compatibility with the luminaires is most important.

Categories of operation (clause 5.6)

The mode of operation and duration shall conform to the results of the risk assessment in accordance with national regulations.

Application standard (BS EN 50172/BS 5266-8)

When the risk assessment is conducted it will give much of the information that is required for a system design. Typical items that will be identified are.

- If the premises are a sleeping risk or if they are not evacuated on supply failure either of these conditions would indicate that the 3 h duration is needed.
- The need for maintained operation of exit signs because many of the occupants may be unfamiliar with the building or the sign needs to be emphasized even if normal supplies are healthy.
- The need for maintained operation of the emergency luminaires because the normal lighting can be dimmed to low levels.
- The presence of hazards which need a fast activation response and high levels of illumination.
- The extent of the escape routes and open areas that need emergency illumination.
- High levels of dirt or long intervals that will reduce light levels and need compensating for in the design.
- The quality of the staff who will conduct the testing and maintenance.

System records and reporting (clause 6)

On completion of the work, drawings of the emergency escape lighting installation shall be provided and retained on the premises. Such drawings should be regularly updated with any subsequent changes to the system. These drawings shall be signed by a competent person to verify that the design meets the requirements of this standard.

Log-book (clause 6)

A log-book must be kept on the premises in the care of a responsible person who has been appointed by the employer or operator of the business and has to be readily available for examination if required by the relevant authority.

It should record at least:

- the date of commissioning of the system;
- the date of each periodic inspection and test;
- the date and brief details of each service/inspection test;

- the dates and brief details of defects and remedial action;
- the date and brief details of any alteration to the system;
- if any automatic testing device is employed, the performance of that device shall be described.

BS 5266-1 provides a template of an ideal typical system log record system. It details action plans to rectify the system and control the extra risks to the building while the emergency lighting is being repaired. To minimize this period adequate spares should be kept for the system. These may include fuses and perhaps a charger controller for the central system. Spares for the luminaires can often best be provided by having complete units which can be readily replaced if necessary.

Servicing and testing (clause 7.1)

When automatic testing devices are used, the information shall be recorded monthly for some systems. This will need to be done manually, either by checking the indicator of each luminaire or from the collated results at a control panel. Other systems will give an automatic print out record of the test results but it is important that any faults that are identified are actioned quickly. For all other systems, the tests shall be carried out and the results recorded. In addition to the test results of the operation of the units, any potential problems that might adversely affect the luminaires should be noted and dealt with, for example a build up of dirt on the diffusers.

Regular servicing is essential. The occupier/owner of the premises must appoint a competent person to supervise the servicing of the system. This person shall be given sufficient authority to ensure that any work necessary to maintain the system in correct operation is carried out.

Testing schedule (clause 7.2)

There is a need to take precautions to safeguard the building during the period of full battery duration tests and while batteries are being recharged. So it is recommended that the tests should be performed preceding a time of low risk should a supply failure occur. Or suitable temporary arrangements should be made until the batteries have fully recharged such as testing alternate luminaires at a time.

Application standard (BS EN 50172/BS 5266-8)

Daily test

The indicators of a central power system should be visually inspected daily. But this could be a time consuming task that could be overlooked, so it is generally preferable to use a repeater indicator and sounder that can be located in a responsible person's office. Provided that the system complies to BS EN 50171, it will have the facility to switch this remote indicator signal and it will also indicate if a fault on the normal supply has activated the system in a remote part of the building. This will alert the responsible person as soon as possible.

Monthly test

A functional test is required monthly. It should be performed by simulating a supply failure so that the changeover device is activated and the lamp is operated from its standby source. After the test the luminaire should be restored to its normal condition and the battery charge indicator should be checked to see that the fitting is back on charge. For central battery systems, any control relays on the system should be checked to see that they work correctly. Generator sets should be tested to their manufacturer's instructions. This will normally involve running them for long enough to reach the working temperature. This is typically about an hour. The test has to be conducted at full load. The fuel should be replenished after this test.

Annual test

A full rated discharge test should be performed to check the battery condition and the lamps operation should be checked. It is ideal if an inspection of the luminaires takes place at the start, in the middle and at the end of the discharge periods as, if it is known that a luminaire has provided half its output and then failed, the urgency of the repair is less than if the condition is unknown as it may not have worked at all.

Central battery systems should be checked to ensure that the cells are balanced in output voltage and that the connectors are clean and tight. If they are vented then it is necessary to check that their electrolyte is topped up and that the specific gravity of the cells is at the correct level.

Allowing electrolyte levels to fall below the plates will permanently damage the area of plate that dries out. After tests the systems should be restored and the charge indicators checked again to ensure that the system is back on charge. Finally, the results of the test should be entered in the log-book.

6. Emergency lighting luminaires (BS EN 60598-2-22)

Luminaires contain the lamp and any necessary control gear to strike the lamp and control its discharge current. In self-contained fittings the battery and charger are within the enclosure, but for centrally supplied luminaires a remote battery and charger system power a number of remote or 'slave' luminaires (see Figure 11).

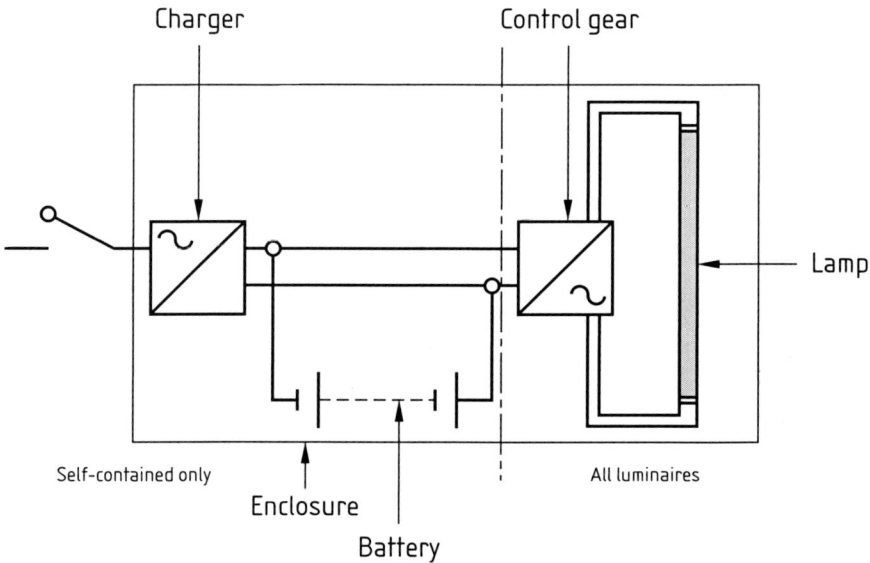

Figure 11 Typical luminaire

Self-contained luminaires

In these fittings the essential components are contained in or adjacent to the luminaire. Some are specifically designed as emergency luminaires, while others are mains luminaires which are converted to have an emergency facility.

The definition in BS EN 60598-2-22 clause 3.8 is 'Self-contained emergency luminaire: Luminaire providing maintained or non-maintained emergency lighting in which all the elements such as the battery, lamp, and control unit and the test and monitoring facilities where provided are contained within the luminaire or adjacent to it (that is within 1 m cable length).'

Centrally supplied luminaires

In these fittings the lamp and some of the control gear is located in the luminaire but the charger, battery and often the changeover device are located remotely and provide the supply to a number of luminaires. Central power units may supply the luminaires with a range of DC voltages or 230 V AC from an inverter. To operate on DC some are specifically designed as emergency luminaires, others are converted mains luminaires. For AC, normal mains luminaires can be used without modification if they are appropriate for the duty.

The definition in BS EN 60598-2-22 clause 3.9 is 'Centrally supplied emergency luminaire: luminaire for maintained or non-maintained operation which is energized from a central emergency power system that is not contained within the luminaire.'

The general test requirements in clause 22.6 define where the distribution cable requirements cease and where the luminaire requirements start. For all luminaires any connecting cables that are shorter than 1 m can be treated as internal wiring. For example, the connecting cable to a projector head does not need special requirements if it is adjacent to the luminaire. If the distance is above 1 m fire resistant cabling should be used and ideally the power supply and light source should not be in separate fire compartments.

This exclusion allows the final drop of centrally powered luminaires to be wired in ordinary cable which allows flexibility if the luminaire is fixed to a suspended ceiling. But in that case the spur connection to the fire protected cable must be fuse protected so that if a fire shorts the

Emergency lighting luminaires (BS EN 60598-2-22)

cable the fuses will rupture and isolate the fault so that other fittings will still be supplied (see Figure 12).

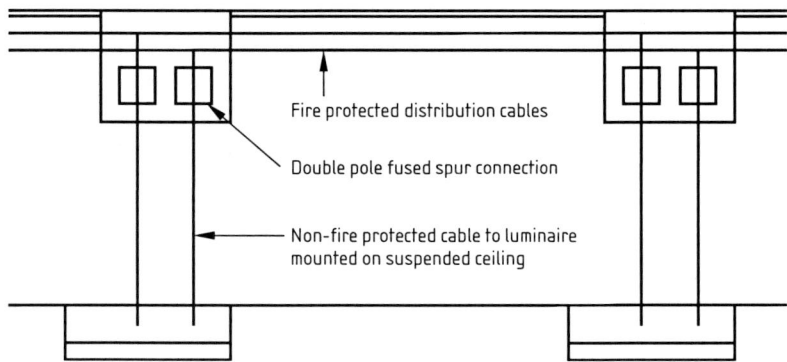

Figure 12 Distribution of power to central system luminaires

If the high frequency inverter is to be mounted remotely then the manufacturer's maximum distances must not be exceeded. This is because the capacitive effect of the cables at the very high frequencies (typically 70 kcs (kilocycles per second)) needed to start most modules absorbs power and prevents starting.

Specific clauses ensure that the product will provide full performance over its design life. The major items covered are listed below.

Chargers (clause 22.19.1)

These must conform to IEC 60924 and BS EN 60598-2-22 (clause 22.19.1). They must provide the rated charge performance specified by the battery manufacturer to charge batteries within 24 h over the rated ambient temperature and operating over the whole range of 0.9–1.1 times the rated supply voltage. While doing this the transformer windings must remain within their maximum permissible temperature.

Transformers (clause 22.19.2)

These must conform to IEC 60742-4.12 and 13. They must have electrically isolated windings (so that there is no chance of the mains supply reaching the battery). The insulation and temperature limits must be satisfactory for a minimum 10-year design life.

Batteries (clause 22.6.8)

These must be designed for a minimum life of four years and they must be applied correctly to provide this life when the luminaire is in operation. The most important factor affecting battery life is cell temperature. This is a combination of the ambient temperature and the heat generated in the cell by the charge current. Exceeding the cell's designed maximum temperature will cause a seriously reduced cell life and can result in catastrophic failure in less time than can be identified within the timing cycle.

Nickel–cadmium batteries can typically operate for 600 duty cycles, which is more than adequate for reasonable operational use. However, it is important that the system is not discharged every night as this would obviously cause premature failures.

For luminaires with three or more nickel–cadmium cells the circuit must provide overdischarge protection to stop one or more cells being driven into reverse during discharges longer than the rated discharge. The gas evolved in reverse is not recombined and the cell becomes permanently and prematurely damaged.

NOTE: Batteries must be labelled with the correct battery disposal method and date of manufacture.

Appendix A of BS EN 60598-2-22 defines the maximum ambient temperature, the charge range and maximum discharge currents of the batteries.

Control gear

This must conform to IEC 60924. The circuit must operate the lamp correctly by providing any preheating required. It should provide a suitable output to meet the lamp supply conditions, to meet the rated output claims, and to enable the lamp to meet the switching cycle tests in the standard.

Changeover circuit (clause 22.17.1)

This must operate at between 0.6 and 0.85 times the rated supply voltage to ensure that the luminaire does not activate under normal

Emergency lighting luminaires (BS EN 60598-2-22)

voltage dips but that it will be fully operational at the level at which most normal luminaires will be ceasing to work correctly.

Lamps (clause 22.5.3)

Luminaires must be clearly marked with the details of the replacement lamps. These details must be visible during lamp renewal. Use of the appropriate lamp is necessary to ensure that the rated luminaire performance and design lamp life is achieved.

Lamp holders (clause 22.5.9)

Luminaires which combine normal and emergency lighting must distinguish the emergency lamp holder by a 5 mm green dot which is visible during re-lamping.

Lamp colour (clause 22.16.4)

In order to identify the safety colours of signs and the coded bands of extinguishers the minimum value of colour rendering index should be better than RA 40. In practice this only restricts the use of low pressure lamps.

Indicator (clause 22.6.7)

In self-contained luminaires there has to be a visible indicator showing that in normal use the battery is being charged. Historically these devices were red in colour. But now that high intensity green LEDs are more readily available, they are gaining precedence. This is reinforced by the need to ensure that if there are automatic test units on a site, then the meaning of the indicator is unambiguous. Consequently a mixture of colours should be avoided, or the operating staff needs to be adequately trained to ensure that they are certain of the signal being given.

Test facility (clause 22.20.1)

Self-contained luminaires have to have the means of testing provided. This can be either integral or remote and may be manual or automatic.

But if it is manual the switch has to be self-resetting or key operated to reduce the chance of either unwanted operation or the supply being left off after a test.

Enclosure

This must be able to protect the emergency circuit and those areas of the enclosure that can come into contact with the DC battery circuit. Either the charger, the batteries or the cables to the ballast must be able to pass the 850°C glow wire test so that any faults within the circuit will be contained by the housing.

The luminaire must conform to at least a rating of IP 20. If higher claims are made, such as IP 54 for external use, these should be tested as part of the approval procedure.

The enclosure must withstand an impact of 0.35 Nm to external parts but higher levels of protection are needed for luminaires exposed to potential attack from vandals (see clause 22.6.4).

In addition to checks on the components the standard also requires normal safety checks and tests of operation and performance.

Abnormal temperatures (clause 22.18)

The luminaires have to be tested to ensure that they will work at 70°C so that, with the luminaires mounted on the ceiling, in a fire condition they should be able to operate for as long as the route is usable. The test checks that at least half of the normal rated duration is available at this temperature.

Photometric performance (clause 22.16)

The tests check the minimum light output and its distribution in 5° steps of gamma in the C planes, C 0 and C 90. This is then used to design systems to conform to BS EN 1838/BS EN 5266-7 which requires that the luminaire rated performance is available through, and to the end of, the rated discharge. The start-up times must also conform as 50% of the light must be provided within 5 s and full output must be provided from 60 s to the end of the rated discharge period. This test identifies if the luminaire exceeds the disability glare limits required in BS EN 1838 and so will need great care in positioning it next to escape routes.

Emergency lighting luminaires (BS EN 60598-2-22)

NOTE: High risk task areas require 100% output to be provided in 0.5 s to the end of the time the hazard is present. This can normally only be met by tungsten lamps or by running the fluorescent lamp in the maintained mode so that the start-up time is avoided as the lamp is already energized.

The data from these tests is de-rated to allow for aging, dirt and lamp life and is processed to form authenticated photometric data either as spacing tables or as part of an appropriate computer program. The data then allows accurate system designs to be developed and used for system design. Verification of the system's compliance is then performed by use of this data, so avoiding the need for difficult measurements being taken onsite.

The requirements for back illuminated exit signs can also be fully tested by checking the values of the minimum patch, the uniformity and contrast to ensure that the sign is appropriately illuminated to ensure good visibility.

7. Centrally powered supply systems (battery systems) (BS EN 50171)

This standard covers central power supply systems that use batteries as an alternative power source. The systems are intended for emergency lighting use but may be suitable for use with other safety equipment such as fire extinguishing installations, paging and safety signalling equipment, smoke extraction equipment and carbon monoxide warning systems. Fire alarm power supplies are excluded from this standard as they should conform to the requirements of their own specific standard BS EN 54. If an uninterruptible power supply (UPS) style of inverter system is to be used for emergency lighting, it must conform to the requirements of this standard in addition to the requirements of BS EN50091-1.

Types of central power supply systems (clause 4)

The central power unit consists of a charger, battery transfer switch and sometimes an inverter and in a supply failure it powers the remote luminaires.

The central system offers a duplicate lighting system which can operate when there is a problem with the normal supply. While they can be produced to operate as non-maintained systems most are able to power the lamps continuously as maintained fittings although in the final area this supply may be held off until needed.

A major advantage of central power supply systems is their ability to be controlled to provide illumination when required. Activation of either emergency or maintained circuits can be initiated either locally or remotely and the system can be inhibited from discharging when the building is unoccupied by interfacing the controls with an essential service.

Where required remote control switching devices (CSDs) can control power to a local area in which the supply to the normal lighting has failed. If the central unit has a maintained output this emergency supply to a specific area can be continued indefinitely until the normal lighting supply is restored. The requirements for safety and operation of these remote CSDs are the same as for the other system components.

Changeover mode (clause 4.1)

In these systems the load is supplied directly from the central unit. When the supply is healthy, the automatic transverse switching device (ATSD) connects the supply to the load if necessary, using a step down transformer to modify the voltage or if a DC supply is required an appropriate rectifier system. In the event of a supply failure the ATSD changes over disconnecting the mains fed supply and connecting a supply from the battery to the load. This output can be modified by the use of an inverter or converter to provided either AC or DC at an appropriate voltage for the luminaires.

The advantage of the maintained output is that final circuit monitoring is not required and if there is a local final circuit failure the luminaires can be continuously supplied as long as the main incoming supply is healthy. In practice, this means that if there is a failure of supply in a local area, the operation of the rest of the premises is not prejudiced.

The maintained output can be interrupted by a link, which enables the circuit to be disconnected when the building is unoccupied. This should be interfaced with an essential service to ensure that it is switched back on when the building is reoccupied.

Mode without interruption (clause 4.2)

It is possible to obtain a maintained output without the use of a changeover device by continuously floating the battery. These systems can also provide AC by the use of an appropriate inverter. While this format does not require a changeover device it requires a significantly larger charger to be able to both supply the load and also recharge the battery. Additionally, this mode of operation requires an expensive charger to limit the load ripple to the battery. If an inverter is connected the battery will also be subject to extensive load ripple. Both of these conditions

Centrally powered supply systems (battery systems) (BS EN 50171)

can seriously reduce battery life and reduce the efficiency of the system, thus incurring excessive energy costs.

Changeover mode with additional CSD (clause 4.3)

If the luminaires connected to either maintained system identified in clauses 4.1 and 4.2 are required to be switched off their supply can be held off by one or more CSDs. The advantage of this system is that the CSD can be located in the area of the final circuit, provided the coil to this device is fed from that final circuit. Failure of that circuit will automatically switch on the emergency lighting as required by BS 5266-1. Additionally the coil circuit can be switched to provide local control of the emergency luminaires while still ensuring that the illumination will be restored on local or total supply failure.

Changeover mode with additional CSD (clause 4.3)

It is also possible to split the load in an area so that only a proportion of it is operating in a maintained mode, the remainder being non-maintained. This procedure is particularly useful to keep exit signs illuminated, while holding off some or all of the other emergency luminaires. This system is particularly useful for cinemas and theatres, where the low levels of illumination needed during performances can then be upgraded in an emergency to ensure speedy evacuation.

Operating conditions (clause 5)

Normal systems are designed to operate on the UK supply voltage with an ambient temperature of up to 25°C and a maximum relative air humility of 85% for locations up to 1000 m above sea level. If the design location requires different values these should be identified as part of the initial design. It should be noted that the battery performance is normally rated at 20°C. Operation below this temperature will reduce the capacity that is available.

To ensure that designs are appropriate the full output that is required and the operational function should be agreed as part of the design brief. This includes the maximum time for luminaires to illuminate and also the load profile, particularly with reference to the starting loads.

Construction (clause 6)

The construction standards are designed to ensure that the equipment will operate satisfactorily and be compatible with luminaires which conform to BS EN 60598-2-22.

To ensure a consistent level of safety, where possible identical standards of requirements have been used for the supply system and the luminaires. For example, the creepage and clearance distances are identical for the same voltages. Where possible appropriate standards have been identified for the components that make up the systems.

Enclosures (clause 6.1)

Enclosures of the main unit and any remote control switch devices must be resistant to fire, have locks and provide adequate physical protection. This is normally achieved by installing the equipment in a suitable cubicle. But systems can be designed for location in a specific locked battery room to provide the necessary protection of the system.

Charger (clause 6.2)

The charger has to be designed to be capable of restoring the battery capacity within the design time period, even under conditions of low mains supply. To demonstrate this after a full discharge of the batteries the input to the charger has to be reduced to the minimum rated voltage and the battery recharged at 20°C for 12 h after which it must be able to supply 80% of its rated capacity.

NOTE: Controlled voltage chargers operate at the maximum current limit until the battery reaches its float voltage at 80% of capacity then the charge current reduces, finally balancing the cell losses.

The charger must withstand an output short-circuit without damage other than to the circuit protection devices, so that if a fault occurs it can quickly be rectified. This ensures that if an output short-circuit inadvertently occurs, replacing the charger fuse or resetting the miniature circuit breaker will restore its output. Consequently, the system will not be out of service while charger repairs are carried out.

Centrally powered supply systems (battery systems) (BS EN 50171)

Transformers (clause 6.3)

Transformers should conform to their own product standard (BS EN 61558-2-6). If transformers are required outside the scope of the standard they should be selected to provide an equivalent level of safety.

Changeover device (clause 6.4)

This device must be rated to the appropriate load category of BS EN 60947. Many emergency loans have a high starting surge and a nonlinear characteristic so the changeover device must be capable of repeated reliable switching operations. Automatic transfer switches (ATSs) must conform to BS EN 60947-4-1 and BS EN 50272-2. They must be rated for the appropriate load that they will need to switch. This must include the starting and switch off loads. Tungsten and fluorescent ballast loads typically take an initial current which is considerably higher than the steady load.

Systems designed to operate on both AC and DC need to be capable of switching both supplies.

NOTE: Particular care needs to be taken with switching high impedance and capacitive loads and high DC voltage switching as the arc is difficult to break particularly at high voltages.

ATSs must change over within the same limits as self-contained luminaires as given in BS EN 60598-2-22. This is designed to ensure that the emergency lights are switched on below 60% of the mains supply, as can occur under brownout conditions. But they do not unnecessarily operate on normal low voltage supply fluctuations down to 85% of mains supply.

NOTE: These limits also apply to any circuit monitors or CDSs (see clause 6.4.1).

Inverters (clause 6.5.2)

The inverter output voltage must be regulated as ±6% of the rated output voltage. This must be maintained for steady loads from 20% to 100% of full load. For instantaneous load changes of up to 10 s this limit is extended to ±10%. This requirement checks that the normal mains luminaires supplied by the system should operate satisfactorily.

Inverter load capability (clause 6.5.3)

Inverters must be capable of supplying an overload of 120% for the system's full rated duration as during the life of the system it is possible that the system will be overloaded. If this is the case it is important that the inverter will continue to operate. The excess load will, of course, reduce the duration available from the battery and this will be able to be identified in the duration test and the excess load can then be removed.

The inverter must be able to start a full load of previously unpowered luminaires in the emergency condition when powered from the battery. Even if the system is an uninterruptible power supply and the luminaires would normally already be on, if there is a distribution fault the voltage depression while clearing the fuse would be likely to shut the luminaires down so they will need to be re-struck to continue operating (see Figure 13).

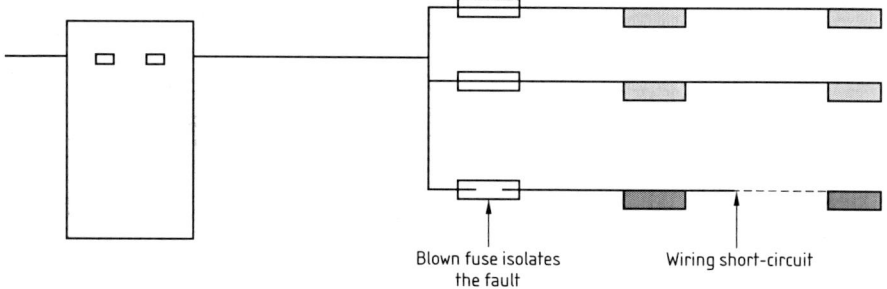

Blown fuse isolates the fault

Wiring short-circuit

NOTE Inverter applications are specified in 6,5,8 of EN 50171:2001 which specifies that inverters must be able to clear any distribution fuses.

Figure 13 Inverter circuits

Fault clearance (clause 6.5.8)

Inverters must be able to clear any final circuit or distribution fuses. If a short on the distribution circuit occurs during an emergency, the system must be able to clear the protective device to isolate the fault and enable the rest of the system to operate. They must be able to perform this test from the inverter powered by the battery, and not from a by-pass

Centrally powered supply systems (battery systems) (BS EN 50171)

supplied from the mains, to ensure that this essential function can be performed in emergency conditions.

Discharge protection (clause 6.6)

Batteries must be protected against the harmful effects of overdischarge. This is achieved by isolating the load after the duty has been performed. The system is automatically reset as the battery is recharged. This system is particularly important for use with recombination lead acid batteries, which are particularly prone to damage if they are left in the overdischarge condition.

Indicators (clause 6.7)

The engineering skills available on most sites are reducing. So, in addition to the essential meters, the control gear must now also be fitted with diagnostic indicators that give an unambiguous signal of the system's condition that does not need technical ability to identify and then take remedial action if needed.

Automatic testing of central power supply systems (clause 6.11)

These systems are optional and should use testing systems conforming to IEC 62034 which is described in Chapter 8.

Centrally powered systems require special consideration when an automatic duration test is used to ensure that the tests are performed at periods of minimum risk which has to be decided a year in advance. Provided that the premises will be empty at the selected test time and for the following 12 h of battery recharge, the standard requirements of IEC 62034 will keep accurate timing that is protected against corruption by supply failures of up to seven days. Unfortunately central power supply systems normally use a single battery so it is not possible to test alternate luminaires while still having a capability to power the remaining fittings. Acceptable procedures are:

- Use of manual initiation of the duration test cycle at the time when the minimum risk can be determined. All the other monthly

function tests and the timing of the test period and checking of the results are then performed automatically.
- Use of special systems with dual batteries and outputs to enable luminaires to be tested from each section of the battery. This is an expensive solution but it does provide maximum security. It also allows battery maintenance and replacement to be carried out while the system is still operational.
- The simplest technique is the use of a limited discharge test. This can only be used because industrial stationary batteries have a characteristic which enables their condition to be accurately assessed by a two-thirds discharge. Typically a 3 h system would be tested for 2 h, and provided that the battery achieved an appropriately higher voltage than the normal 3 h discharge voltage, the battery can be shown to be in good condition.

NOTE: These procedures can also be used for manual testing.

Batteries (clause 6.12)

These must be designed for industrial duty and to accept a continuous charge to maintain full capacity.

- Battery life. The cells used must have a minimum life expectancy of at least 10 years except for small power packs (up to 1500 Wh^{-1}) which must have a 5-year expectancy. Car batteries are specifically excluded. Their plate construction is not designed for constant charge current and they would fail quickly and without warning.
- The batteries must be designed to be able to provide the full system rated output at the end of their design life.
- The batteries must not be discharged to a voltage less than 90% of the system's nominal voltage.
- The batteries must be protected by a low voltage disconnection circuit. This protects them from overdischarge which could be harmful to their life expectancy.

Battery safety is important as the high amounts of stored energy need to be treated with respect. So, the appropriate clauses of BS EN 50272-2, the European safety standard for battery installations, are given below.

Centrally powered supply systems (battery systems) (BS EN 50171)

- BS EN 50272-2 (clause 5.1). Protection against both direct and indirect contact. The requirements for the safety categories of HD 384 must be complied with. For systems up to 60 V nominal the requirements for extra low voltage apply. For voltages above that and up to 120 V nominal there are specific safety conditions, including that they shall be located in a locked enclosure.
- BS EN 50272-2 (clause 7.2). During maintenance there must be a minimum distance of 1.5 m between the touchable live parts of any batteries above 120 V. For battery systems above 120 V insulated protective clothing and local insulated floor coverings will be required.
- BS EN 50272-2 (clause 8.2). The ventilation requirements must be met. Although many systems now use recombination batteries that do not require topping up, the issue of adequate ventilation can become important under end of life fault conditions. While the natural ventilation is adequate in most installations, if additional ventilation is required, it is desirable to use fans on the inlet side of the ventilation system, rather than the extract side, which under battery full conditions could have high concentrations of hydrogen.

8. Automatic test systems for emergency lighting (IEC 62034 and other standards)

Testing of emergency lighting is essential to provide confidence that the system will operate correctly when an emergency occurs. Unfortunately the action is time consuming and is often neglected, particularly if the manual facilities are difficult to use. This is an ideal task to be automated as it takes away the drudgery of a repetitive job that often just proves that the system is working correctly. To be acceptable, automatic testing of the emergency lighting must be at least as effective and as safe to operate as manual testing.

The major requirements are that:

- The changeover device has to operate correctly and supply the lamp from the battery during the test.
- The tests must be performed at the correct intervals for the required times.
- The result of the test must be reliably indicated.
- Facilities are provided for the test to be done at times of low risk.

The different forms of automatic testing systems are defined and classified by type reference so that the mode of operation is clearly understood.

Self-contained with stand alone facilities (type S)

These units have an initiation circuit, timer, testing control and result indicator built into the self-contained luminaire. They operate independently of other luminaires and require no additional wiring or control panel. Consequently they are the easiest form of automatic testing to install. While they initiate the tests and indicate the result they still need to be visited once a month to log the result of the test. Because of this they are mostly used for small to medium premises or where there is already a safety patrol who can readily add the recording to their other duties.

Self-contained with remote panel (type P or EN)

By connecting the luminaires to a control panel it can gather the test results at a single point, thus reducing the task of recording results (type P), or the panel can be fitted with printer facilities which produce a hard copy that is acceptable as the log record (type EN). In these systems the initiation, timing and monitoring circuits are normally in the panel although they can remain in the luminaire and the panel just repeats the test results. Many of these systems offer additional functions. Apart from showing the identification of any faulty luminaires they can even diagnose the component that has failed, enabling service engineers to obtain the correct spares to rectify the luminaire.

It is also possible to show the location and fault on a mimic diagram that can be printed out to direct service staff straight to the fitting.

Central powered system with remote panel (type P or EN)

Central systems have to operate with a panel although this can be built into the power unit cubicle because the panel has to switch the change-over device to the luminaires. As the full system has to be tested and full load must be applied to the batteries any hold-off or remote control relays have to be activated to power all the emergency luminaires. The timing and monitoring circuits and indicators have to meet the same requirements as the self-contained system and they are also designated as (type P) without printer and (type EN) with printer.

General requirements (clause 4.1)

Manufacturers must identify the classification of the system they are supplying and also the type of luminaires that are suitable to operate with it.

Automatic test systems for emergency lighting (IEC 62034 and other standards)

The safety and constructional requirements of the test circuits are the same as those used for the luminaires (specified in BS EN 60598-2-22) and for the centrally powered systems (specified in BS EN 50171) so that when they are integrated into products, the protection provided is consistent.

To ensure that the indicated results can be relied on, the test duration periods and their frequency of operation can only be reset by authorized personnel.

Timing circuit (clause 4.2)

The system relies on the timing functioning and also on being accurate. If the timing stopped the user could record the next monthly results not aware that another test had not been performed. To prevent this situation the timer has to monitor its own function, typically by use of a watchdog circuit, so that if the timer fails a fault signal is given.

Test function (clause 4.3.1)

The system has to check that each component of the emergency circuit is operational and any faults such as the charge fail indicator must be reported within 24 h. Consequently, a manager sitting next to the panel will be aware of any luminaire or central power unit failures shortly after they occur and so is able to take remedial action quickly.

Emergency supply (clauses 4.3.2–4.3.4)

During the test period the monitoring circuit has to indicate if the emergency supply has failed. This is particularly useful for converted fittings or centrally supplied luminaires with one lamp being switched from the mains to the emergency circuit and the others on the mains supply where it can be difficult to see visually if a change has been made. In all systems the test circuit has to check that all the changeover devices have operated correctly. Particular requirements are detailed for those luminaires which provide a maintained output without a changeover by continuously powering the inverter circuit when the lamp has to be checked in both in the normal mains healthy and in the mains failed conditions.

Protection against component and intercommunication faults (clause 4.4)

Any fault of components or interconnecting wiring faults must not be able to inhibit the operation of the emergency circuits or induce an unwanted test which would leave the system discharged. This ensures that any open circuit or short-circuit faults or line shorting to earth will not affect the operation of the emergency system. Any of these faults must be identified within or during the monthly functional tests.

As with any system, great care must be taken to ensure that the individual components are compatible. For example, many systems monitor lamp current and these need to cover the range of the lamps used, also the number of luminaires and their location zones need to be within the capabilities of any monitoring panels. Electromagnetic compatibility (EMC) legislation has to be complied with, particularly for emissions and immunity from corruption by any other equipment. Many of the interconnections carry sophisticated digital data and the correct transfer of this information must be able to be relied on.

Testing of lamps (clause 4.5)

During the test the lamps have to be checked to ensure that they are connected and also operate. This is straightforward for tungsten lamps but fluorescents have to be checked to ensure that current is flowing through the lamp and not just the heaters. The circuit is checked by the test house connecting equivalent rating lamp heaters to the lamp holders instead of a working fluorescent lamp. The test system must then indicate that the lamp is faulty.

Functional test (clause 5.1)

This test simulates a supply failure and activates the changeover circuit to power the control gear and lamps from the battery. The monitoring circuit checks that the lamp is operating and gives a positive signal

Automatic test systems for emergency lighting (IEC 62034 and other standards)

through the indicator to show that the test was satisfactory. Each of these tests are just for sufficient time to strike the lamp and should not exceed 10% of the rated emergency duration of the system to minimize the discharge from the battery and hence the duration that would be available if an emergency occurred directly after this functional test. By performing this test the operator of the building can be confident that the luminaire works. The only item not checked is the length of time for which the battery will run the fitting. These function tests should be performed at least once a month to detect luminaire failures as quickly as possible with minimum damage to those components that are affected by the test, in particular, the battery and lamps.

Duration tests (clause 5.2)

To check the battery capacity available a full discharge is needed. This should be done at least once a year. Provided that good quality approved equipment is correctly used this testing regime will identify the gradual reduction in capacity that occurs prior to sudden loss of power. Obviously, while the test is being performed and the battery is being recharged the luminaires will not be able to provide their designed duration of illumination. Consequently arrangements have to be made to enable systems to perform this test at times of minimum risk. When commissioning, this test is performed at the start of the timing cycle to check that the system is working correctly.

Protection of the occupants of a building during test and recharge periods (clause 6)

Because the duration test is set one year in advance a number of precautions are necessary to limit the risks that this necessary test imposes. They depend on the facilities provided by the system and whether the building is likely to be occupied or not. The main conditions are given below.

Unoccupied buildings

If it is known that the building will be unoccupied for the test and subsequent 24 h recharge, it is possible to test all the luminaires at once. It is possible to set the time for the duration test a year ahead.

The system requirements to perform the test then are:

- It must be possible to be able to set the scheduled test for that date.
- The timer must be sufficiently accurate to do the test at the right time.
- The timer must be able to maintain its timing even for extended outages of supply failure.

Possibly occupied buildings: self-contained luminaires

To protect occupants alternate luminaires have to be tested and recharged, after which the remainder can be tested. This procedure ensures that no areas of the building will be in total darkness, as at least two luminaires have to illuminate each compartment of the escape route or open areas.

The system requirements to perform the test then are:

- The system must be capable of being set so that alternate luminaires can be duration tested at times that are at least 24 h apart to maintain the same timing accuracy and protection as is needed for buildings that will be empty.
- The timer must be sufficiently accurate to do the test at the right time.
- The timer must be able to maintain its timing even for extended outages of supply failure.

Possibly occupied buildings: central systems

As there is often only a single battery available it is not possible to test the battery by checking alternate luminaires, but fortunately these systems use stationary batteries which reflect their state of charge by their output voltage. Consequently systems should be discharged for two-thirds of the rated duration. If by that time the battery voltage has not sunk below a preset figure for the battery type, we can be confident that the full capacity will be available to the lower rated end of the discharge voltage.

This technique ensures that at least one-third of duration is always available even immediately after the test.

Automatic test systems for emergency lighting (IEC 62034 and other standards)

Alternatively it is acceptable to provide dual batteries which can be individually tested while the other is available to power the luminaires. This system gives a high level of security as it also provides cover while batteries are replaced at the end of their life but unfortunately it is costly.

Buildings known to be unoccupied 24 h in advance

For some premises such as town halls, which are hired out, it is impossible to predict in advance when they will be unoccupied. So it is permissible to allow manual initiation of the duration test at a safe time. The timing of the test monitoring and indications of the results must then be automatic.

Timing accuracy (clause 6.2.1)

The timer must have an accuracy of better than 75 s per week over the full temperature range that it is likely to encounter. In practice the temperature tends to normally be fairly constant so the accuracy is going to be improved. But even so the worst case is that a test could vary by up to 2 h for each year. This means that when nickel–cadmium batteries are reaching replacement time the total timing variation will be only a maximum of 8 h by which time the battery will have recharged and have at least 2 h duration available. The timer shall have an accuracy of better than 75 s per week.

Protection of the timing function (clause 6.2.2)

The systems rely on an accurate timing circuit that is immune to mains supply failures to perform the tests correctly and at the right times. The timing device must not be affected by extended power supply failures of up to seven days. This facility means that if there is a lengthy supply failure, or if the operator inadvertently switches off the supply over a weekend, then the timing will still be maintained. The user will then not be forced to reset the programmed times of the test or worse still, if they neglect this, have the tests being performed at a time of potentially high risk. The manufacturer's instructions have to make it clear that if there has been a supply loss of longer than seven days the system should be re-commissioned, including the full duration test. This is particularly important in applications which are only used seasonally, such as holiday establishments, when it is important to ensure that the premises are safe before opening to visitors.

Indication and recording of tests (clause 7)

It is important that the indication of the test results is logged each month, if they are not printed automatically, but the equipment must ensure that any fault signals are not overridden by a positive signal to a lesser test. For example, if a luminaire fails the full rated discharge test, perhaps because there is only 1 h available from a 3 h rated fitting, this signal must not be cancelled by a satisfactory result one month later to a functional duration test of 5 min.

There should be at least one system condition indicator if the test control gear is installed in a self-contained luminaire. It can be combined with the charge health indicator, but in all cases the colour coding of IEC 60073 should be followed. This means that healthy conditions should be shown by a green lamp and fault signals by yellow or orange. Dangerous conditions are shown by a red indicator. Operating staff should be conversant with the indications and any other information provided by the system and be able to rectify any faults identified.

Selection of a suitable system

The most useful type of system depends on the format of the building it will be used in, the way it will be operated and the type of emergency lighting it will work with.

It may be difficult to install the wiring needed for a panel controlled system in an existing building and self-contained independent luminaires may be preferable. Some advanced versions of these systems use hand-held beam communicators so that the diagnostic data given on panels can be downloaded and a full record made of the system condition.

Testing modules are available that can be fitted in existing luminaires but generally it is worth installing new fittings as the benefits of fitting the installation tend to outweigh the savings from retaining old luminaires.

New buildings enable any system to be installed easily and matched to appropriate emergency luminaires for the application.

If there are no suitable maintenance staff available a panel system with printer provides a full report and in the event of a failure

Automatic test systems for emergency lighting (IEC 62034 and other standards)

diagnostic systems can identify the component needing to be replaced as well a site location.

All automatic testing systems reduce the testing burden as the test does not have to be initiated and staff do not need to be trained to either perform the test and interpret the results or be monitored to see that they are keeping to the work schedule. Automatic testing gives a reliable monitoring of the system condition and early warning of any faults that may affect the emergency lighting system.

9. Other relevant standards

Emergency lighting also requires the use of a number of application, specialist and component product standards to design satisfactory systems. These are listed below.

BS 5266-2

This is the code of practice for electrical low mounted way guidance systems for emergency use. The use of way finding techniques can be a useful addition to an emergency lighting scheme particularly if there is a chance that smoke may penetrate the escape route. The standard details the amount of light appropriate for each low mounted marker and how far apart they should be mounted for optimum visibility. It also specifies how doors should be identified as to whether or not they are for use as exits.

BS 5266-4

This is the code of practice for design, installation, maintenance and use of optical systems.

Optical fibres offer a number of advantages in some applications, notably that the illumination source can be located in a readily accessible area and the optical light guide is then used to distribute the light where it is required. The systems are also used for aesthetic reasons and for convenience of installation, for example as low level indicators of stairs in low ambient lighting applications such as cinemas and theatres. In addition, because the lighting distribution is non-electrical, these systems are useful in explosive atmospheres and those areas where no electrical interference can be tolerated such as radio communications areas. The standard details the design and installation practices including how to calculate the transmission losses in the optical cable.

BS 5266-5

This specification details the quality and protection that is necessary for the components for optical systems, including the fire resistance and the procedures to ensure that a single lamp failure does not invalidate the system.

BS 5266-6

This is the code of practice for non-electrical low mounted way guidance systems for emergency use: photoluminescent systems. Photoluminescent systems use paint which is charged by the presence of light and then discharges with a glowing effect if the lighting fails. These systems are also a useful addition to an emergency lighting scheme when used to offer a way finding system particularly as they can be mounted at a low level to assist in locating the escape route in the event of smoke. As this equipment needs no electrical supply it is easier to install than powered systems but it can only be used in areas that receive adequate illumination prior to possible use. The standard details the amount of material that is required to give an adequate intensity of output and the procedure as to how to test the material that has been installed to prove that it still provides the design output.

Battery standards

BS EN 50272-2

This gives essential detail as to the safety requirements for the use of secondary batteries and battery installations in central systems. It defines safe working levels and practices that should be adopted to protect against the electrical dangers for systems up to 60 V nominal, for those up to 120 V and for those systems above that figure. It also provides the ventilation calculation which should be used because even recombination cells can vent in fault conditions.

Other relevant standards

BS EN 60285

Sealed nickel–cadmium cylindrical rechargeable cells are the batteries used in most self-contained luminaires and this standard not only defines their physical size but also the testing regime to ensure the cells are suitable for the ambient temperature, charge rate and discharge current for the design life of four years required for emergency lighting.

BS EN 60896

Stationary lead acid cells are the types of cells that can be used in centrally powered systems. Part 1 of this standard covers the vented types that need topping up of the electrolyte. Part 2 details the valve regulated types which do not require topping up and consequently are the most popular for emergency lighting applications.

BS EN 60623

This standard covers vented nickel–cadmium cells which are capable of operating under arduous conditions and over a much wider temperature range than equivalent lead acid batteries. They can be left discharged without any ill effects and also have a long service life expectancy.

Luminaire standards

BS EN 60598-1

This is the overall standard for all luminaires and has many parts for specific applications (like BS 60598-2-22 for emergency luminaires). This part 1 contains many of the test procedures that ensure luminaire quality and is referenced whenever possible by the emergency luminaire section to avoid constant repetition of the requirements.

BS EN 61347-2

This standard gives the requirements for electronic ballasts for fluorescent lamps. It details the control gear for emergency luminaires and as many of the tests of performance as possible are contained in it in order to reduce the unnecessary duplication of tests of complete products.

Lighting terms and photometry

BS EN 12665

This document was produced by the CEN 169 Working Group and contains the definitions and derived formulas for all lighting applications. It provides the base information that is used by all the application specific standards and details the template for the common use of computer programs.

BS EN 13032

This is the document for emergency lighting applications. In addition to specific definitions for the applications, it details the factors that should be used for the data presentation to conform to BS EN 1838/ BS 5266-7.

Wiring systems

BS 7671 Wiring systems (IEE Regulations)

This is the UK version of EN HD 364 and it selects those sections which are applicable in the UK. The reference to emergency circuits is in section 5-56 and while this is being reviewed the principle will remain that self-contained luminaires are treated like any other luminaire. Central systems require procedures to ensure that the supply from the central power unit to the luminaires is protected in the event of fire.

Light and lighting mains lighting for indoor workplaces (BS EN 12464)

This standard was produced by CEN 169 Working Group 2 and details the appropriate minimum lighting levels for different applications. It also has suitable values for colour, uniformity and glare. It is relevant to emergency lighting as it details the normal lighting conditions that will be in place before the emergency occurs. This is particularly crucial in high risk task area applications.

Other relevant standards

Light and lighting: sports lighting (BS EN 12193)

This standard was produced by CEN 169 Working Group 4 and details the lighting requirements for sports stadia. It is of specific interest for emergency lighting applications as it details the values of lighting that must be provided for protection of the participants in dangerous sports over and above the emergency lighting levels for evacuation. These levels are expressed as a percentage (typically 5–10%) of the normal lighting values and are only required to be provided for the time that the hazard exists (typically 30–120 s). Sports areas that should be covered, that are in common usage, include those for swimming and indoor gymnastics. As these requirements are for covering against the dangers of sudden supply failure, the lighting is required to be available within 0.5 s but there is no need for the circuits to be protected against fire.

Emergency lighting (CIE 5-19)

This standard is based on the work of CEN 169 Working Group 3 that is published as BS EN 1838/BS 5266-7 but it also includes guidance on protection of escape routes in the event of smoke entering them.

Other relevant documents

These include:

- CIBSE Emergency lighting design guide. This has some very useful advice, particularly on the de-rating factors that should be applied for lighting design calculations;
- ICEL 1001 ICEL Industry trade association scheme for registration of fittings with third-party approval and authenticated spacing table to meet BS EN 1838/BS 5266-7;
- ICEL 1006 ICEL Application design guide for emergency lighting;
- ICEL 1004 ICEL Scheme for registering of acceptable conversion of normal luminaires for emergency use.

10. Regulatory Reform (Fire Safety) Order

Legislative background

The Fire Precautions Act required that premises with more than 20 people employed or six people in sleeping accommodation must have a fire certificate.

The guidance on how to comply with the Workplace Directive originally endorsed the use of The Fire Precautions Act and demonstrated compliance by issuing fire certificates. But in addition, it extended the scope to all premises with employees by requiring employers to conduct and act on a risk assessment.

The legislative situation was confused and difficult to comply with or monitor because there were many separate elements of legislation, much was overlapping and some even seemed contradictory.

The Regulatory Reform (Fire Safety) Order amends or replaces 118 pieces of legislation, the most significant being the repeal of the Fire Precautions Act 1971 and the revocation of Fire Precautions (Workplace) Regulation 1997. Anyone familiar with the 1997 Regulations will recognize much of the detail that is in the order; as it develops and extends many of the concepts from them. The order has reduced the use of fire certificates to a few high risk premises. In its place employers will be expected to provide a fire safety risk self-assessment which the fire authority will be able to audit to check that it conforms to the appropriate guidance for the application.

Regulatory Reform (Fire Safety) Order

The order was added to the statute books in June 2005 for implementation in April 2006. It applies to the majority of premises and workplaces in the

UK. But it does not apply to: dwellings, the underground parts of mines, anything that floats, flies or runs on wheels, offshore installations, building sites or military establishments.

The order firmly places a responsibility on the responsible person and outlines all the measures that must taken to ensure the safety of all the people he or she is directly or indirectly responsible for. At the same time it allows the enforcing authority to make sure that it is enacted (by force if necessary) and sets penalties if it is not.

It requires the responsible person to carry out a fire risk assessment, produce a policy, develop procedures (particularly with regard to evacuation), provide staff training and carry out fire drills.

The responsible person must provide and maintain clear means of escape, emergency lighting, signs, fire detection, alarm and extinguishers.

The specific requirements of the order are contained in the following sections:

- the employer must provide means of escape [4. (1) (b)] and ensure that they are available at all times [4. (1) (c)];
- signs and notices indicating emergency routes and exits [114 (2) (g)];
- fire detection and alarm. An appropriate fire detection and alarm system must be provided [13 (1) (a), 4. (1) (e) and 15 (2) (a) and (b)]. The type and extent of the fire alarm are subject to the requirements of the risk assessment;
- emergency lighting: escape routes must be provided with emergency lighting [114 (2) (h)].

The 'responsible person' is normally the employer but it may be the manager, occupier or the owner of the premises. The requirements define that empty buildings are now brought within the scope of the order.

The responsible person must produce a risk assessment for the premises and make any necessary changes to their premises to ensure that the risk is at a tolerable level.

To assist them the government has produced guidance documents for 11 specific application categories. Fortunately, for emergency lighting the guidance is broadly similar for all categories.

Regulatory Reform (Fire Safety) Order

Provisions for persons responsible for compliance with the regulations

The responsible person must ensure contractors appointed or employed to design, install, maintain or test fire safety or test equipment or systems are 'competent'. The definitions of competent vary slightly but in principle they say that the person has to be someone who has the necessary knowledge, training experience and abilities to carry out the work.

For emergency lighting this means that the engineers must be familiar with and able to design to the latest issues of the relevant standards. This in turn means that they must be able to meet the photometric requirements of the standards. They also need access to the full product data on the equipment they intend to use.

They are not only responsible for the safety of employees but for anyone on the premises or who might be affected by a fire and for liaising with other employers who may be within the same building.

Fire safety risk assessment

Risk has two components balanced against each other: the possibility of an occurrence and the consequences of that occurrence. For example a metal fabrication workshop has a high possibility of a fire due to the cutting and welding equipment. But providing the house keeping is good and no combustible substances are present, then a fire is not likely to spread so the consequence is low, therefore the risk can be considered to be normal.

BSI has produced PAS 79:2005 to give guidance on the methodology of fire safety risk assessment. But to be able to understand the implication of the technique for emergency lighting system design it is important to understand the basic principles.

The Regulatory Reform (Fire Safety) Order uses the same techniques as have been established for health and safety and requires employers to:

- carry out a fire risk assessment: for five or more employees, a record must be kept;
- monitor and review assessment: revise as appropriate;
- inform staff or their representatives of the risks;
- plan for an emergency and provide staff training;

- provide and maintain: to the extent that it is appropriate, determined by the assessment:
  - means for detecting and giving warning in case of fire;
  - means of escape and emergency lighting.

The order will not contain specific fire safety requirements but a condition requiring compliance with the new order.

Where the enforcing authority is not satisfied that the precautions are adequate, rather than stipulate what is required, they will advise where, in their opinion, the precautions do not conform to the law.

In a dispute it would be for the accused to prove their innocence and that all practicable steps have been taken. In effect, the burden of fire safety is being shifted from the fire brigades and licensing authorities across to the 'responsible person' within the private sector.

The 'responsible person' can be the employer, manager, occupier or the owner of the premises. This means that empty buildings are brought within the scope of the new order.

Provisions for persons who are responsible for compliance to the regulations

The responsible person must ensure that the contractors appointed or employed to design, install, maintain or test fire safety or test equipment or systems are competent.

They are not only responsible for the safety of employees but for anyone on the premises or who might be affected by a fire.

The principle of the regulations and the risk assessment approach is goal based and flexible to employers' needs. The employer generates the risks in the workplace; therefore, to safeguard the safety of employees, the employer must.

- Identify hazards and people at risk.
- Remove or reduce the hazards.
- Manage the remaining risks to acceptable levels by:
  - ensuring that all occupants are alerted and can leave the premises safely in the event of a fire;
  - reducing the probability of a fire starting;
  - limiting the effects should a fire occur.

STEP 1 Identify fire hazard: sources of ignition, fuel, work processes.
STEP 2 Identify the location of people at risk.
STEP 3 Evaluate the risks. This includes checking that the existing fire safety measures are adequate, that the fire alarm system will provide warning and that the emergency lighting will illuminate the escape routes and signage to enable the occupants to vacate the premises. The maintenance and testing procedures for these systems must be checked and be acceptable.
STEP 4 Record findings and action taken. Prepare emergency plan and train employees
STEP 5 Keep assessment under review and revise if situation changes.

Anything that burns is fuel for a fire. So the responsible person needs to look for things that will burn reasonably easily and are in sufficient quantity to provide fuel for a fire or cause it to spread to another fuel source. Some of the most common 'fuels' found are:

- flammable liquids: paints, petrol and adhesives;
- flammable chemicals;
- wood, wood shavings and offcuts;
- textiles;
- packaging materials;
- synthetic wall and ceiling coverings like polystyrene tiles.

The main source of oxygen for a fire is in the surrounding air. In an enclosed building this is provided by the ventilation system in use. This generally falls into one of two categories: either natural airflow through doors, windows etc., or mechanical air conditioning and air handling systems. There is normally a combination of systems, which will introduce/extract air to and from the building. Additional sources of oxygen can sometimes be found in materials used or stored in a workplace such as chemicals (oxidizing materials).

If there is a fire, the priority is to ensure that everyone reaches a place of safety quickly.

Putting the fire out is secondary. The greatest danger from fire is the spread of the fire, heat and smoke.

If a workplace does not have adequate means of detecting and giving warning or means of escape, a fire can trap people or they may be overcome by the heat and smoke before they can evacuate.

As part of the assessment, those at risk in a fire should be identified, and how they will be warned and escape should be considered. To do this employers need to consider who might be at risk, including customers, and ensure they are protected by adequate emergency lighting.

Additional fire protection is necessary if the occupants of the buildings have specific problems. They may be disabled and thus require assistance to escape (places of refuge should have emergency lighting and facilities to summon help).

Elderly people have much worse sight at low light levels than young occupants so if there is a high proportion of the elderly either high levels of contrast are needed at changes of direction, and particularly for the colours of stair nosing, or higher levels of illumination should be provided. If any staff are working at a remote location or they are working by themselves they are at a greater risk than those in populated areas. Adequate illumination is needed for them to be able to access places of safety.

If staff are performing maintenance tasks in areas that are not normally occupied, care must be taken to ensure they will be warned and have reliable temporary illumination.

Fire risk categories for assessing the means of escape

Appropriate safeguards such as emergency lighting are needed to reduce risks to tolerable levels by compliance with BS 5266-1. The risk categories are defined below.

High risk

This category of risk is caused by risks from structure or contents, where there is a risk to life in case of fire, and where there are difficulties for evacuation. Examples of risks from structures or contents include: an unsatisfactory structure, for instance a lack of fire-resisting separation, complex escape routes, the storage or use of flammable or explosive materials, large areas of flammable or smoke-making surfaces, and work activities with potential for fires, e.g. kitchens.

A significant risk to life in the case of a fire can occur in sleeping accommodation and care areas.

High risks during an evacuation can include: a high proportion of elderly or infirm people; groups of people in isolated areas; large numbers of people present relative to the size of premises, and only a low level of assistance being available.

Normal risk

This occurs where fire would spread slowly and few people are at risk.

Low risk

This category covers situations where there is minimal risk to life, the risk of fire occurring is low, and the potential for fire or smoke spreading is negligible.

Reduction of high risks

Additional emergency lighting beyond current levels may be needed. The employer's assessment of fire hazards will identify a high risk when there is both a high possibility of the occurrence and a serious consequence from that occurrence.

Increasing the areas covered by emergency lighting safeguards the use of the escape routes from mains failure.

In areas of high hazard the escape routes should be through those parts of the area with minimum risk.

As far as possible the system should be protected from the effects of a likely fire.

Other health and safety risk assessments, such as those controlling hazardous machinery, should be checked for the need of high risk task lighting.

Reduction of normal risks

The current standards should be adequate in situations where: an outbreak of fire is likely to remain confined or spread slowly, allowing people to escape to a place of safety; where the number of people present is small and the layout of the workplace means that they are likely to escape to a place of safety without assistance; and where the workplace has an effective automatic warning system.

Fire hazards or obstructions in a means of escape

The following items should not be in escape routes:

- portable heaters of any type or any oil or gas cylinder fuelled heaters or cooking appliances;
- upholstered furniture;
- temporarily stored items;
- lighting using naked flames, e.g. gas boilers, pipes, meters (except those permitted in the Gas Safety Regulations);
- gaming and/or vending machines;
- electrical equipment (other than normal lighting, emergency escape lighting, fire alarm systems, or security systems), e.g. photocopiers.

High levels of emergency lighting mitigate this problem.

In many locations occupants are funnelled through narrow gaps. The risk assessment will show that higher levels of illumination are needed. Examples include:

- checkouts in supermarkets;
- turnstiles at sports venues;
- ticket barriers at railway stations;
- immigration barriers at airports.

Language difficulties can also create communication problems.

To meet safety requirements a new building must conform to the Building Regulations and a risk assessment must be available to demonstrate that the way in which the building will be used is adequate to meet the Workplace directive and the Regulatory Reform (Fire Safety) Order. Any local authority requirements must also be met.

The guide specifies that emergency lighting installations should be installed by competent engineers and they should conform to BS 5266-1. Significantly no date is given, which means that it must be to the latest edition (1999).

Major changes that have occurred with different issues are that in 1988 0.2 lx minimum for corridors and 1 lx average fall open areas was specified. In the 1999 version the requirement was changed to 1 lx minimum for most corridors and 0.5 lx minimum for open areas.

The guide gives very stringent conditions for escape routes by defining minimum widths and a long list of items that should not be on the route.

As part of the risk assessment the employer now has to face the problems of testing and repair as they are fully responsible for the operation of the equipment.

Points they should consider when purchasing the units are:

- The quality of the units. This reflects the service life that can be expected so they should look for confirmation that the product complies to the relevant product standard and that it has a Kitemark or equivalent national mark. This demonstrates that the product has been third-party tested. Kitemark registration also checks that the manufacturer is operating an ISO 9000 series quality assurance program to ensure build quality and performance is as good as the initially tested product.
- ICEL registration, in addition to checking that the product is Kitemarked, provides third-party approval that the photometric spacing data provided meets the requirements for use as authenticated data when used in designs.
- Ease of testing, both functional and duration.
- Ease and cost of maintenance, battery/lamp change.
- Likely support from manufacturer during the life of product.

The risk assessment should identify the areas of a premises that present risk to the occupants. That risk must be minimized to tolerable levels by the use of appropriate fire precautions. Emergency lighting to BS 5266-1 is deemed to meet that requirement for normal risks. Other areas or higher levels of protection are likely to be needed for high risks. For instance, the presence of many people in an open area would require emergency lighting even if the room was below 60 m^2 in floor space. To remain legal the employer must test and maintain the installation to BS 5266-1.

11. The Building Regulations

Approved Document B

The safety of the building structure is detailed in the Building Regulations Approved Document B which details the fire safety requirements that are required. Compliance is checked by building control officers during the building process of new buildings and also for major refurbishments. This activity complements the role of the fire authority who then ensure that the building is used safely by the occupants.

The document controls the materials used to provide fire compartments and also ensures that appropriate numbers and size of exit routes are available. Included within this section is the requirement that adequate emergency lighting to BS 5266-1 must be provided to ensure occupants can cross open areas and then use the escape routes provided.

Table 9 of the document defines the areas requiring protection for specific purpose groups of the building. Clause 6.36 requires that the areas specified in Table 9 shall have adequate emergency lighting installed to BS 5266-1.

To assist in understanding what types of buildings are defined as which purpose group Table D1 gives examples of applications for each classification of the purpose groups. Unfortunately the numbering given against the group numbers of the sections of Table 9 do not correspond to the group references in Table D1 so care needs to be taken to ensure that the correct purpose group names are used.

Other requirements identified by the risk assessment should have identified areas that may have a particular risk that needs extra protection. This may be because of storage of flammable materials, the presence of sources of ignition or the activities of the occupants.

A Guide to Emergency Lighting

BS 5266 recommends consultation between the building control officer, user, designer and fire authority to agree the precautions needed. This is particularly necessary in premises with multiple occupancies to ensure compatibility of the systems.

Approved Document B details those parts of a premises that should conform to BS 5266-1. In addition a risk assessment may identify specific hazards that also require protection and in some cases local authority license conditions will need to be observed.

The building should be checked to ensure that appropriate escape routes and open areas are covered.

A checklist is useful to assist in evaluating whether the emergency lighting is adequate for a new or an existing building. A concise version is given in Annex C of this book.

Emergency lighting compliance checklist

Check of system type and documentation

1.1 Are the correct areas of the premises covered? Check with risk assessment and Approved Document B.
1.2 Is the system documentation correct and available? This must include completion declaration sheets from: the designer detailing the performance of the system; the installer to confirm the design has been met; and the verifier to demonstrate that the requirements have been achieved.
1.3 Has the system been designed for the correct mode of operation category? Check with risk assessment.
1.4 Has the system been designed for the correct emergency duration? Check with risk assessment.
1.5 Is the completion certificate with photometric design data available?
   The photometric design data should consist of either authenticated spacing tables for the luminaire used or a computer print of a suitable program that details that valid data has been used and that the required factors to allow for the worst case at any time in the rated duration and at the end of design life with aged lamps and with the effects of dirt on the diffuser.
   Alternatively adequately de-rated actual test results of the installation can be used.
1.6 Is a test log available and are the entries up to date? To assist the system designers and installers to demonstrate that their

The Building Regulations

system is operative they need to perform the tests and record the findings and take any appropriate action. A typical test log system is given in BS 5266-1:2005. Specific test requirements may also be advised by the equipment manufacturer. To be able to keep a system operative proper test procedures and records must be kept. These should start with the commissioning test. Then at least monthly function tests of operation are needed. Annual full duration tests should also be made. The results should be recorded and any failures logged in an action plan until rectified. It is also recommended that appropriate action to safeguard the premises is planned at the design stage. This should include the storing of appropriate spares to speed repairs.

Luminaires

2.1 Are the luminaires installed those documented in the design? Check designations. Cost pressures and sometimes honest mistakes can result in a lower performance luminaire being used on the premises. It is important to ensure that if an alternative has been used the original design requirements are still met. The spacing coverage of a well engineered 3-cell, 8 W luminaire can often be double that of a 2-cell 8 W fitting. The size of a lamp is little guide to its performance.

2.2 Are the exit signs and arrow directions correct? The location and operation of exit signs is vital to a speedy evacuation particularly when many of the occupants may be unfamiliar with the building. If there is any uncertainty as to the escape route additional signs should be used.

NOTE: The directional arrow should show that the route is straight through an exit door. If another direction is to be taken when through the door, that should be signed appropriately.

2.3 Are there luminaires sited at the 'points of emphasis'? While navigating the escape route occupants need to be able to negotiate hazard points such as changes of level and also to locate and operate fire alarm call points and extinguishers. By locating an emergency light near to these positions it both acts as a beacon and provides the highest illumination on the route.

2.4 Is the spacing between luminaires compliant to spacing tables or drawing? After the hazard points are located the minimum illumination of all points on the route has to be verified. If the spacing seems excessive a visual inspection of the illumination should be made. If concerns remain the locations should be checked against verified data. If in doubt the spacing table for the luminaire should be consulted.

2.5 Is there illumination from at least two luminaires in each compartment? This requirement is not to affect the light level but to stop a section of the escape route being plunged into total darkness in a supply failure if a single luminaire has failed. If the doors have vision panels to the next compartment this is acceptable. The requirement only applies to compartments of the escape route so the lobby of a toilet does not normally need to be covered provided that when one door is held open the next door can be reached.

2.6 Are the luminaire housings suitable for their location? The luminaires should be adequately robust, vandal resistant and weatherproof to ensure that they will not be damaged in their service life. As a general guide polycarbonate housings will take considerable impact and manufacturers' claims for a fitting to be vandal resistant are checked by third-party test houses. Weatherproof protection is also tested and the IP rating system is used, generally outdoor locations would need IP 43 or better.

2.7 Are non-maintained luminaires monitoring the local lighting circuit? To ensure that the emergency lights illuminate even if the normal lighting failure is caused by the final circuit of the lighting circuit, non-maintained fittings have to be operated from that supply. Self-contained luminaires can be connected directly but central systems normally use a remote sensing relay to activate the changeover contactor. Checking this function is normally performed by requesting the final lighting device to be operated.

Compliance with BS EN 60598-2-22

3.1 Do the emergency luminaires conform to BS EN 60598-2-22? Compliance with this product standard confirms that the product will provide the required output for the required duration and meet the designed component life. Use of a luminaire with known output characteristics enables designs to be checked and verified without extended and difficult onsite acceptance tests. Evidence of third-party testing from BSI or other approved test house declares that the tests have been met and that the production complies to ISO 9000 rules. In addition ICEL registration also checks the spacing data provided for the unit to assist system design. If these standards are not marked on the luminaire other evidence of compliance should be given, particularly with regard to the quality and life to be expected from the battery.

3.2 If a central power supply unit is used does it conform to BS EN 50171? Central power units are vital to the performance of the system. Compliance with the product standard ensures that the battery and charger are of a good quality and compatible with each other. Also that the changeover device is capable of switching the operational and fault loads. For AC output systems the inverter is checked to see that it is able to supply and is compatible with the load.

3.3 For centrally powered systems is the wiring fire resistant?

For central supply systems the integrity of the supply between power unit and luminaires must be maintained even through fire conditions. This is achieved by: siting the power unit in a low fire risk area; using cables with a high grade of fire resistance and ensuring that any junction /distribution boxes maintain the cable fire resistance; using luminaires that are protected so overload faults will not affect other luminaires.

Wiring that must maintain circuit integrity during the course of a fire (primarily wiring between a central battery unit and the associated slave luminaires) and must either be inherently fire-resisting (e.g. mineral insulated copper sheathed (MICS) cable), or be protected against fire by burial in the building structure. The BS 5266-1 Completion certificate requires a declaration of compliance by the designer and system installer.

3.4 Do any converted luminaires conform to BS EN 60598-2-22? Use of normal mains luminaire housings for emergency lighting applications is desirable aesthetically to many users, however, converting fittings has to be correctly performed to achieve BS EN 60598-2-22. The control gear and battery must be approved to BS EN 60695. The temperature of the battery in the fitting should be under 50°C. And EMC standards must be maintained. In addition to thermal and electrical safety tests converted luminaires have to be reassessed for EMC compliance and the converter has to take responsibility for the resulting fitting.

Test facilities

4.1 Do the test facilities simulate a supply failure? The test must simulate fully the conditions that can occur in a supply failure in addition to checking that the non-maintained luminaires are activated by the failure of adjacent mains lighting. Central batteries must be tested on full load, any hold off relays must be bypassed and any dimmed circuits checked to see they do not inhibit full emergency lighting output.

4.2 Are the test facilities safe to operate and do not isolate a required service? The test facility must be appropriate – a retractive push button can be ideal for a function test but how will the duration test be operated? Withdrawal of live fuses is not a safe procedure and over time arcing could degrade contact. They are safe to use to ensure isolation when the circuit has been switched off. Care must be taken that the test switch does not also supply another important function such as mains lighting in a stairwell where interruption would be dangerous.

4.3 Are the test facilities clearly marked with their function? Inadvertent operation of a test switch will at very least discharge the battery leaving it unable to operate. Consequently, in addition to being located where they will only be operated by competent engineers, they should also be clearly marked as 'Emergency lighting' and the marking should also indicate its use.

4.4 Are the user's staff trained and able to operate them and record correctly? Operators must be trained to understand and keep to the test schedule intervals. They must know the procedure to initiate and terminate the test when checking that recharge is commencing. They must be able to assess if the test is satisfactory and note any needs for maintenance such as blackening of tube ends or excess dirt on diffusers. Most importantly, they must be able to imitate rectification in the event of the failure of a test.

Escape lighting

5.1 Are escape lighting cables segregated from all other cables? BS 7671, BS 5266-1.8 and 6B.5 all make it clear that emergency lighting circuits are to be segregated from other normal supplies. Under fault conditions one supply could also take out the other and segregation increases the chance of one supply remaining healthy. Also if a supply got onto the other system it would be both dangerous and damaging.

5.2 Are all isolators, switches and protective devices kept to a minimum? BS 7671 and BS 5266-1.8 identify that joints, isolators and protective devices all reduce the integrity of the distribution system and so they should be minimized. There are benefits from some distribution fusing to limit the area of loss if a fault occurs so it is generally accepted that a maximum of 20 luminaires should be fed from a single fused supply.

5.3 Can any AC systems start the lamps from the battery in an emergency? In a supply failure it cannot be relied on that the luminaires will already by operating. Fluorescent and tungsten lamps both have considerable inrush starting currents – typically these can be up to 10 times their full load consumption. While this only occurs for a fairly short time period inverter switching and control circuits have to be designed to provide this load. Compliance should be checked by performing the test with the supply to the inverter switched off.

5.4 Can any AC systems blow all distribution fuses/MCBs in an emergency? Another area of concern on some systems is their inability to clear faults by not being able to provide the power to operate the distribution over current protection. Obviously in a supply failure condition with a fire, if a circuit suffers a short the system must be able to isolate that faulty area and continue to supply the rest of the building.

12. System design

Design objectives

To create an optimum design for the premises it is important that all aspects of its application should be considered. Emergency lighting is an essential part of the building services of premises. Normally the main reason for its installation is to protect people in the event of fire by assisting them to evacuate the building safely. However, it also reduces the risks to occupants if they are plunged into darkness during a power failure, and helps to protect property from theft. These subsidiary benefits should also be taken into account in the design.

Initial considerations

To ensure the system is well designed and as reliable as possible, planning is important through all phases of the project, from considering legal requirements to final commissioning and maintenance. Consultation between all the interested parties at an early stage of the design is important to avoid expensive modifications to the completed system.

The people who should be consulted on most installations are considered below.

The employer or operator of the premises as defined within the new legislation. They should have a 'responsible person' who will have conducted a risk assessment of the premises. This should have identified the people at risk and the particular hazards that will need protecting. They can also give useful input as to the competence of their own staff to use, maintain and test the systems being considered.

The information they should be able to provide in addition to the risk assessment is current plans of the building, details of associated fire alarm and extinguisher provision. If the premises are already in operation, details of previous fire certificates and the history of any fire incidents is also useful.

The system design engineers should have outline performance data and the costs of likely systems and be able to give advice on the advantages and disadvantages of their use. They should be familiar with the legal requirements for the type of premises being considered. They should have access to adequate photometric design information for whatever type of system is selected.

The fire authority should be consulted to ensure that the system is being designed to current legislative requirements, and that any particular hazards are being adequately considered.

The developing role of the fire authorities will be to audit the responsible person's risk assessment. This will include checking that the emergency lighting system is appropriate to limit the risks within the building and also that maintenance and testing schedules are adequate and carried out.

The building control inspector will check that the building regulations are complied with for new installations and major refurbishments. Fortunately, compliance with the requirements of the regulations is likely to include those items that will be identified by the risk assessment. The inspector will need to be satisfied with the fire safety provisions before they sign off the building as being safe for use.

The installing engineer will need to be familiar with the installation regulations and capable of meeting the manufacturer's requirements for the system that will be selected. They will need access to drawings showing the distribution of supply to the normal lighting luminaires to ensure all that non-maintained emergency luminaires will operate on a local final circuit failure. If drawings or records are not available this information must be found by inspection of the system.

The manufacturers of the emergency lighting equipment will be needed to advise the design engineers of the performance of their equipment. They need to be able to confirm the photometric performance, either in the form of authenticated spacing tables or of a suitable computer design program, in addition to any specific application details that should be considered.

System design

Legislative requirements

The first stage of system design is to gather the information needed on the project, normally by consultation with the regulatory authority and the user. This should cover legislative and likely operational requirements, and customer preferences. Because legislation and guidance is constantly evolving it is important that the latest edition of documents (directives, standards, guidance notes etc.) should always be referred to.

There is a considerable amount of British and European legislation affecting emergency lighting. The major items are given below.

The Construction Products Directive (89/106)

Section 4.3.8.1 of this directive defines: 'Emergency lighting installation (panic lighting, escape lighting). The purpose of the installation is to ensure that lighting is provided promptly, automatically and for a suitable time in a specific area when normal power supply to the lighting fails. The purpose of the installation is to ensure that:

- the means of escape can be safely and effectively used;
- activities in particularly hazardous workplaces can be safely terminated;
- emergency actions can be effectively carried out at appropriate locations in the workplace.'

The Construction Products Directive covers both buildings and civil engineering works including domestic, commercial, industrial, agricultural, educational and recreational buildings, as well as roads and highways, bridges, docks and tunnels. It requires that such buildings or works are designed and built in such a way that they do not present unacceptable risks of accidents in service or in operation such as stumbling or tripping in poor visibility, and that the safety of occupants and rescue workers is ensured in the case of fire. Minimum standards of illumination are required so that people may move safely within the works, including if they have to escape. In addition, escape routes are required to provide secure and adequate lighting, capable of operating despite failure of the electrical supply.

In the UK this is implemented by the building control officers and applies to most new and refurbished buildings except for private dwellings.

The Workplace Directive (89/654)

This legislation ensures that premises are safe for the activities conducted by the occupants. It has a number of safety requirements, including:

Clause 4.5 Specific emergency routes and exits must be indicated by signs in accordance with the national regulations.

Clause 4.7 Emergency routes and exits requiring illumination must be provided with emergency lighting of adequate intensity in case the lighting fails.

In the UK this legislation is being implemented by the fire authority and 11 new guidance documents are being issued by the ODPM to cover specific applications. They will clarify that compliance requires that the user must produce a risk assessment for all premises in which people are employed. The fire authority will then be able to audit these assessments to check that they are adequate and that the fire safety precautions such as emergency lighting are adequate.

The Signs Directive (90/664)

This is implemented in the UK by Statutory Instrument 341. The relevant clauses are:

- Clause 6 Depending on requirements, signs and signalling devices must be regularly cleaned, maintained, checked, repaired, and replaced.
- Clause 8 Signs requiring some form of power must be provided with a guaranteed supply.

The Safety Signs Directive is retrospective and was implemented in the UK on 1 April 1996. It calls for the provision of emergency signs in all places of work. These signs must be regularly cleaned, tested and maintained, and visible at all times. The traditional text 'EXIT' signs must have been replaced by the pictogram by December 1998. In the UK the Health and Safety Executive have passed responsibility for ensuring compliance to the fire authority, and have produced a combined

System design

guidance document covering the use of safety signs. It is called 'A guide to Statutory Instrument No. 341, The Health and Safety (Safety Signs and Signals) Regulations 1996', and has been published by the Health and Safety Executive (L64).

Other UK legislative requirements

Some workplaces require a licence or registration from the local authority. All these premises require compliance with BS 5266, but in addition the fire authority may require higher levels in specific areas.

Predesign information

The initial considerations should provide system designers with a brief of application information and requirements to be met by the installation.

Before designing the emergency lighting scheme the following information needs to be determined to ensure the proposed system is appropriate.

- The duration of the emergency lighting: 3 h duration is required in places of entertainment and for sleeping risk. Three-hour duration is also required if evacuation is not immediate, or early re-occupation may occur. Normally most are designed for 3 h duration as the cost penalties are small in comparison with the flexibility that this additional time period allows in the use of the premises. A major advantage of this practice is that if the legal requirement is only for a 1 h duty then this is the period that the equipment should be tested for. However, it is important that the test engineers understand that if a 3 h design luminaire fails to provide 1 h duty then replacement is a matter of greater urgency as the battery will have degraded further than a failure at the 3 h rate.
- Emergency lighting of the maintained type should be used in areas in which the normal lighting can be dimmed and in areas where a build-up of smoke could reduce the effectiveness of normal lighting. Maintained lighting which combines both emergency and normal lighting functions may also be desirable for aesthetic or economic reasons. Maintained lighting does not

need to be initiated by the local final circuit failures and because the lamp has been continuously energized it is more suitable for applications requiring a fast provision of the emergency supply such as high risk task areas.
- The exit signs always need to be illuminated to be visible at all times when the premises are occupied. Because of the difficulties of ensuring that the normal lighting will adequately do this maintained signs are required in licensed and entertainment venues and they should be used in any premises which are used by people who are unfamiliar with its layout. This consideration now applies to applications such as shops, where in the event of a bomb threat, occupants may suddenly need to know the location of the nearest exit. If the sign is illuminated even though the normal supply is healthy its message is emphasized.
- Building plans need to be obtained showing the location of the fire alarm call point positions, the positions of fire fighting equipment, and fire and safety signs. The design needs to illuminate these positions so their location must be accurately known.
- Emergency escape routes should be established, and potential hazards investigated. The risk assessment should identify both of these items but it is important that the significance of the loss of normal lighting is considered and the risks caused by it minimized.
- Open areas larger than 60 m^2 floor area or areas identified by the risk assessment that require emergency lighting. Buildings produced before the 2000 edition of the building regulations may not have covered rooms above this size limit. Also the risk assessment could also identify smaller rooms, which still required emergency lighting and their use should be identified and their protection agreed.
- High risk task areas should be identified. Because the emergency lighting is 10% of the normal lighting that level has to be established. The provision of this higher level of emergency lighting has to be provided within 0.5 s and remain for the period that the hazard is present so this time also needs to be established. If the hazard is some heavy duty machinery, such as a large milling machine, then the hazard only applies while the machine is slowing down and stopping in the event of supply failure. For these types of risks 5 or 10 min would be an acceptable duration for the high emergency level. If, however, the risk were an open acid bath, then the hazard would be present for the full emergency escape duration and the high risk task lighting should

System design

match the duration of the emergency escape lighting in the area. In addition to the time correction the area to be covered should also be agreed with the requirement being that the vicinity of the hazard is protected. This area should be agreed with all the interested parties.

- Determine the need for external illumination outside final exit doors and on a route to a place of safety. Occupants are not regarded as safe until they are away from the influence of the building, and at a place of safety. The location of that place and the route to it should be identified in the premises emergency plan and it should be agreed if luminaires just sited at final exit points will provide sufficient illumination. In some cases, with the agreement of the fire authority, it may be possible that adjacent street lighting may fulfil this requirement. If this decision is made it should be identified in the documentation supplied with the completion certificate.
- Other areas that need illumination, although not part of the escape route, should be located, e.g. lifts, moving stairways and walkways, plant rooms and toilet accommodation over 8 m^2 gross area. These areas are not parts of the escape route, but are areas that could need to be evacuated or visited in a supply failure. Consequently, emergency lighting is required for the protection of the visitors and all staff in those areas.
- For central systems, a low fire risk location for the battery units and cable runs should be established. Ideally, the location of the central battery units and the power supply should be separated from the normal lighting supply, so that a fault to one system would not affect the other one.
- For non-maintained applications the area covered by the final circuit of the normal lighting has to be determined, as self-contained luminaires must be fed from that final circuit and it must be monitored by the central system.
- Standby lighting requirements should be established, if activities need to continue during a failure of the normal lighting supply. While to provide true standby lighting full normal lighting levels are needed there can be premises where lower levels can be used to accomplish some tasks. For example, over supermarket checkouts a good level of emergency illumination may enable customers to be checked out in the event of a supply failure, assuming that there is no need for emergency evacuation. Such an arrangement has safety benefits as in an emergency the higher level of illumination will assist customers to negotiate the

A Guide to Emergency Lighting

constricted aisles. In a normal supply failure the ability to check out customers in an orderly fashion reduces their frustration as they can complete their purchases and the operator is not faced with the problem of having to return half-full trolley loads of goods back to their shelves.
- The customer's preference and operating considerations should be ascertained as they will have to operate and maintain the system. They need to be capable and trained to understand the interface of all the safety systems and they need to be appropriate for their operating practices.
- Appropriate testing systems and maintenance procedures must be determined also considering the capabilities of the staff and to establish when competent engineers should be called in. In a village hall it may be appropriate to test the system by isolating the supply. However, in an old people's home breaking the total supply would be potentially dangerous as well as disturbing for the residents. Any testing switches must not isolate any other essential supply and they must be safe to use as well as clearly marked and should not be operable other than by authorized staff.
- Any hazards or locations of people at risk identified by the risk assessment should be covered to limit the risks to a minimal or negligible level.

Design of new installations

Systems should be designed to meet BS 5266-1:1999 and requirements of European and draft European standards.

Design objectives

When the supply to any part of the normal lighting fails, an emergency lighting system meeting the requirements of BS 5266 and BS EN 1838 apply and escape lighting is required to fulfil the following functions.

- Show the escape routes clearly and unambiguously.
- Provide illumination along such routes to allow safe movement towards and through the exits.
- Ensure that fire alarm call points and fire fighting equipment provided along escape routes can be readily located.
- Allow operations concerned with safety measures to continue.

System design

Stage 1 design procedure

Initially luminaires should be sited to emphasize the position of difficult areas on the escape route, and also to highlight the location of safety signs, fire alarm call points, control panel and fire extinguishers. The luminaires act as beacons over parts of the escape route that may be dangerous at low levels of illumination and also highlight other safety equipment that may need to be operated.

This procedure should be performed regardless of what part of the building is considered and whether the area is an emergency escape route or defined as an open area.

The term 'near' in the requirements means that the luminaire must be sited within 2 m of horizontal distance from the specified location. It can be either wall or ceiling mounted depending upon the application. These locations are mandatory requirements of BS 5266-7/ BS EN 1838 and typically cover most of the luminaires needed on escape routes.

Near stairs or any other change of level

Changes of level are generally the most difficult area of the escape route to negotiate, particularly at low light levels, so it is important that these areas are adequately protected. To assist safe passage the luminaires must be located so each tread receives direct light. Generally at least two luminaires will be needed to provide the 1 lx minimum level on the centre of each tread. (Even old designs engineered to the original value of 0.2 lx needed the higher level on the treads unless contrasting colour stair nosings were fitted.)

It is difficult to obtain precise spacing for stairways, so it is generally safer to work on the spacing that is a minimum over the mounting height range from the top to the bottom of the stairs.

Going up stairs towards a luminaire the calculation of light levels has to accommodate the varying mounting height of the luminaires from each tread. The spacing from a luminaire is often reduced as the height decreases as the points illuminated rise up the stairs and the cosine correction factor reduces the light.

Conversely going down stairs the spacing from the luminaire may be reduced as the cosine correction improves in comparison with the floor level as the treads descend. At some point the effect of increased distance from the luminaire will outweigh this.

Other changes of level, such as stages and ramps that can cause tripping hazards in low light levels must also be illuminated.

A Guide to Emergency Lighting

Near changes of direction and intersections of corridors

At any position that the escape route changes direction, or if it intersects a corridor, the luminaires act as beacons to indicate the route and also provide the most illumination where two streams of escaping occupants could be joining. Generally, the best use of illumination from the luminaire is to have the luminaire sited at the intersection so that light shines along the corridor and the intersected route.

Outside and near to exits

Weatherproof luminaires should be located outside final exits and consideration should be given to using fittings which are appropriately vandal resistant. These luminaires are needed, because the safety of occupants must be protected until they are away from the influence of the building. If the area outside the building has hazards in the darkness, such as a river bank, the risk assessment should determine if further emergency luminaires are needed until a place of safety can be reached.

If street lighting is available and adequate it may be used with the agreement of the fire authority.

Near fire fighting equipment and call points

The luminaire must be sited within 2 m (measured horizontally) of any extinguishers, hose reels, fire alarm control or repeater panels and fire call points. By locating the luminaire in proximity to this fire safety equipment it acts as a beacon directing the eye to the safety equipment. It also ensures that the fire equipment, which will have instructions for its safe use, receives the maximum illumination by being located under the luminaire. This measure is particularly important now that the body of all extinguishers is red and the type coding is limited to a small coloured band and at low light levels the eye's colour recognition performances is poor.

Near first aid post

This category was introduced in BS 5266-1:1999 and recognizes that if the normal lighting supply fails but there is no fire requiring immediate evacuation, then access to and use of other safety equipment must be maintained. This measure covers the area of the first aid station to a

System design

minimum of 0.5 lx. If, however, there are facilities to cover complicated procedures, higher levels of illumination should be used.

Illuminated exit and other safety signs

To be clearly visible signs should be mounted as close as possible to the occupant's line of sight. For this reason, regulations state that signs should be located between 2 and 2.5 m above the floor. If this is not possible the fire authority should be consulted to ensure that the meaning of the sign is still acceptable and the reasons should then be identified as a deviation in the system completion documentation.

While this normally relates to exit direction and first aid signs the risk assessment may indicate that other safety signs may need to be visible during a supply failure such as a radioactive warning, so these would also need emergency illumination.

The output of exit signs should not be used in the photometric calculations unless their characteristic has been tested and authenticated data is available.

To enable the sign to be seen and its message recognized BS 5266-7/BS EN 1838 specifies the minimum illumination of any patch of the sign, its uniformity and the contrast between the white and green colours. These values are difficult to measure or verify but signs that are tested and conform to BS EN 60598-2-22 will have been checked as part of their test procedure. If the sign relies on remote illumination, this can be checked by a suitable computer photometric design program. If neither of these procedures is possible it is important that the sign should be visually checked in the emergency condition to ensure that it is adequately conspicuous. It is also essential to check that the emergency luminaire is within 2 m of the sign and that the sign receives direct illumination from the lamp. Many installations, particularly those with large open areas, use interim painted hanging pendant signs which can be moved to accommodate changing room layouts. Unfortunately they are often located too far from an emergency light, or the emergency light may have a louvered distribution, which will prevent the face of the sign receiving sufficient illumination to enable it to be recognized.

BS 5266 and BS EN 1838 state that:

> 'Signs are required at all exits, emergency exits and escape routes, such that the position of any exit or route to it is easily recognized and followed in an emergency. Where direct sight of an exit or emergency exit is not possible and doubt may exist as to its

position, a directional sign (or series of signs) should be provided, placed such that a person moving towards it will be progressed towards an exit or emergency exit.'

The format of signs

BS 2560 used the old format of sign with only green words out of a white background.

These signs should all have been replaced by 24 December 1998 but some are still in existence.

The BS 5499-1 format has the addition of a pictogram of a running man in a doorway with the word 'Exit' and a directional arrow.

The European format uses a full pictogram only showing a man running towards a door. It was implemented as a legal requirement in the UK by Statutory Instrument 1996 No. 341 on 1 April 1996.

The format of the sign and its use is defined in the HSE guidance document site (L64).

The guides for compliance with the Fire Safety Order being issued by the ODPM clarify that either BS 5499-1 or the European format are acceptable but they should not be mixed on a single site.

Exit and safety signs: maximum viewing distances

It is important that the exit signs are of sufficient size to be able to clearly identify the route. At the maximum distance the rectangular green panel should be visible and as the occupants progress towards it the detailed information such as the direction of the arrow becomes visible. Maximum viewing distances are given in EN 1838 as 200 × the height of the pictogram for internally illuminated signs. Because they do not stand out so obviously, signs which rely on external illumination have to be twice the size of the self-illuminated sign (100 × the height of the pictogram). If the distance would result in signs being too large, it may be desirable to use intermediate repeating signs.

Illumination requirements for safety signs

To enable the sign to be seen and its message recognized BS 5266-7/ BS EN 1838 specifies the minimum illumination of any patch of the sign, its uniformity and the contrast between the white and green

System design

colours. BS EN 1838-5 details the emergency illumination conditions for a sign to be clearly visible for the distances specified above.

- The colours must conform to ISO 3864 (white figures with green background for exit and first aid signs).
- The minimum luminance of any part of the signboard is 2 cdm^{-2}.
- The ratio of maximum to minimum luminance of any area of either colour of the sign shall not be greater than 10:1.
- The ratio of luminance between white and the colour shall be between 5:1 and 10:1.

These values are difficult to measure or verify but fortunately signs that are tested and conform to BS EN 60598-2-22 will have been checked as part of their test procedure. If the sign relies on remote illumination the values can be checked by a suitable computer photometric design program. If neither of these procedures are possible it is important that the sign should be visually checked in the emergency condition to ensure that it is adequately conspicuous. It is also essential to check that the emergency luminaire is within 2 m of the sign and that the sign receives direct illumination from the lamp. Many installations, particularly those with large open areas, use interim painted hanging pendant signs which can be moved to accommodate changing room layouts. Unfortunately, they are often located too far from an emergency light or the emergency light may have a louvered distribution, which will prevent the face of the sign receiving sufficient illumination to enable it to be recognized.

The output of exit signs should not be used in the photometric calculations for the design of escape routes or open areas has unless their characteristics have been tested and authenticated data is available.

If the exit door or has a crash bar or other facilities that require operation it is desirable to use an exit sign with a panel providing downward threshold lighting.

Additional areas requiring emergency lighting

In addition to covering the escape routes and open areas the standards also require that emergency lighting is provided to cover areas that may be a risk in the event of a supply failure or that staff may need to enter them to ensure the safety of the building.

Consequently additional emergency lighting should be provided at these locations:

- Lift cars. Although these are only part of the escape route in exceptional circumstances, they may present a problem if the public are trapped in them in the event of a supply failure. Emergency lighting enables the occupants to locate and operate the alarm bell to summon assistance. In addition, the presence of illumination reassures those who are nervous about their trapped condition. The product standard for lifts requires emergency lighting to be fitted, but it is wise to ensure that it is tested and operates satisfactorily.
- Motor generator, control or plant rooms require battery supplied emergency lighting to help any maintenance or operating personnel who may have to access in these areas in a supply failure. This also includes the winding rooms of lifts, to enable trapped staff to be released.
- Toilet facilities and other similar areas such as swimming pool changing areas exceeding 8 m^2 floor area or with no borrowed light and all toilets for the disabled should be provided with emergency lighting as though they were open areas. The latest edition of BS 5266-1 clarifies that this requirement does not include the bathroom within a hotel suite. Where this room has a small modesty lobby before the escape route is reached the assessment may conclude that, provided the door can be reached and opened before the toilet door is shut, the lobby does not require its own emergency luminaire. If this decision is appropriate it should be identified on the deviation section of the completion certificate.
- Escalators should not be used in a supply failure as the treads will be of uneven height and there is a chance that they could restart taking people back towards the fire. However, emergency lighting must be provided for the protection of those using the escalator in the event of a supply failure and, realistically, if a stationary escalator was the only local way out it would be used regardless of any notice to the contrary.
- The pedestrian escape routes from covered and multi-storey car parks need to be identified and provided with emergency lighting. Open ground car parks do not require illumination unless a risk assessment identifies a particular hazard.

System design

Illuminance requirements for escape routes

When the points of emphasis have been identified a suitable luminaire should be selected and its light output performance data obtained either in the form of a spacing table or as a computer design program.

In addition to luminaires at the points of emphasis, it may be necessary to provide extra luminaires to ensure that minimum light (illuminance) levels are met along the whole escape route. For 2 m wide escape routes, the illuminance is specified along the centre line with 50% of that illuminance over the 1 m wide central band. Wider routes should be treated as open areas or as multiple routes.

Illuminance levels for escape routes

The European standard (BS EN 1838) requires 1 lx along the centre line of escape routes and this is acceptable for all normal applications including those with temporary obstructions such as hotel trolleys. However, so as not to render some existing installations non-compliant, the UK has a national 'A' deviation which still allows us to accept the old value of 0.2 lx along the centre line. However, this condition can only apply for installations with 'permanently unobstructed' escape routes. During discussions in Europe regarding the best light level, it was realized that, even if during normal operations an escape route could be kept clear, an obstruction could occur during the emergency period. Consequently, BS 5266 now recommends that all installations should provide the minimum of 1 lx on the centre line. In addition to this recommendation, during any refurbishments it is wise to consider updating to the higher level as the 'A' deviation could be removed at any time. Thus, it is sensible to meet the 1 lx level at any time when changes are being made to the system.

With experience it is possible to mentally evaluate the optimum spacing for any interim luminaires that may be needed. Until this experience is reached, the distance between points of emphasis should be noted and divided by typical spacings to find the best balance between the number of luminaires and the cost of obtaining that performance. When making this evaluation the cost of both product and installation should be considered.

During the photometric design of a system it is important to check that the light level is met with all smoke control doors shut as this will be the site condition that will apply during a fire alarm condition.

A Guide to Emergency Lighting

If the escape route is wider than 2 m it must be treated as a number of 2 m wide strips.

In addition to meeting the light level, the number of luminaires may be influenced by the need to have illumination from at least two fittings in any compartment of the escape route. This requirement is designed to avoid the possibility of escape times being significantly extended by the need to pass through a section of total darkness. However, the condition is met either by there being an illuminated exit sign in addition to a single luminaire or if there is illumination from a visible panel in the door to the next compartment.

BS 5266 and BS EN 50172 recommend using a larger number of low power luminaires rather than a few high power units. Each compartment of the escape route should be lit by at least two luminaires thus, if a luminaire fails, the route will not be plunged into darkness.

Escape routes: BS EN 1838/BS 5266-7 requirements

For escape routes up to 2 m in width, the horizontal illuminances on the floor along the centre line of an escape route shall not be less than 1 lx and the central band consisting of not less than half of the width of the route shall be illuminated to a minimum of 50% of that value.

Illuminance levels for open areas

Open areas require emergency lighting if they are larger than 60 m^2, have an escape route passing through them, or if a hazard has been identified in the room by the risk assessment. An example of an area that could be smaller than 60 m^2 but could be considered as a risk requiring emergency lighting would be a school chemistry laboratory where students handling acids would be at risk if plunged into darkness.

In these cases emergency illumination is needed from at least two luminaires to prevent the occupants being subjected to total darkness in the event of a single luminaire failure.

These areas were previously required to meet an illumination level of 1 lx average but unfortunately, the increasing use of louvred and luminaires and tungsten–halogen projectors meant that these simple design criteria no longer provided safe installations. The requirements could be met by a few high pools of light, with the remainder of the floor being in a relative darkness. Consequently, the new standard had to be

System design

developed to ensure that a minimum level was available over the total area likely to be used. This was particularly important as during the life of the installation the layout of rooms and escape routes through them would be likely to be changed a number of times.

Open areas: BS EN 1838/BS 5266-7 requirements

The standard requires 0.5 lx minimum anywhere in the central core of the floor area. This core area excludes the 0.5 m to the perimeter of the area.

The requirement relates to the empty core area, so the shadowing effects of movable objects in the core area are excluded but fixed items such as structural pillars are included.

Spacing tables

Authenticated spacing tables provide the information to help decide how many extra fittings are needed to meet the minimum light levels in addition to those required for the points of emphasis. BSI or an equivalent test house have to produce photometric tests on approved luminaires giving the lighting distribution round the fitting and the initial and end of life total light outputs. From this data the manufacturers construct tables to allow easy design for installers. The accuracy of the tables is independently verified by ICEL under an ISO 9000 approved scheme.

Verified values are given for fittings registered to ICEL 1001 as authenticated data that has been derived from BSI test data de-rated for the end of battery and lamp design life with allowance for the effects of dirt and ignoring reflection. These are the conditions to give the worst case values required to be used for system design to conform to BS EN 1838/BS 5266-7.

The authenticated spacing tables (see Figure 14) show the distance from the wall or door to the first fitting and then the distance that must not be exceeded for spacing between subsequent fittings. This is shown for the fittings being mounted either parallel to the route (axial) or at right angles to the route (transverse) for different mounting heights. In addition to values for escape routes figures are also given for the coverage of open areas by regular arrays of luminaires.

A Guide to Emergency Lighting

Ceiling mounting height m.	Escape Routes 1 l_x minimum along centre line				Open areas 0.5 l_x minimum in central core			
	Transverse to wall	Transverse to transverse	Axial to axial	Axial to wall	Transverse to wall	Transverse to transverse	Axial to axial	Axial to wall
2.5	2.7	7.2	4.6	1.5	2.5	8.5	7.4	2.3
4	2.1	7.5	4.8	1.7	2.6	9.8	8.6	2.3
6	-	5.3	3.7	-	1.8	10.3	9.5	1.6

Use of authenticated spacing tables if the transverse to axial spacing is needed. Add one half of the transverse to transverse to one half of the axial to axial value.

Figure 14 Spacing table for typical luminaire

Changes in light level requirements

Major changes to the lighting requirements were made in the 1999 issue of BS 5266-7/BS EN 1838. If systems were installed to the old requirements, they need to be re-assessed to check if they still provide appropriate levels of safety. The changes in light levels are shown in Table 1.

The completion certificate requires that appropriate photometric calculations accompany the documentation. This can be in the form of either:

- authenticated spacing tables, such as those produced and registered for ICEL 1001 for the products used on the site;
- first principle calculations as detailed in CIBSE/SLL Guide LG12;
- appropriate computer print out of design results.

As this documentation is a necessary part of producing a system that complies with BS 5266-1, it is logical to check that a verified photometric performance is available before commencing system design using a specific luminaire.

System design

Table 1 Summary of changes to illuminance requirements

	Current: BS EN 1838/ BS 5266-7:1999	**Old: BS 5266-1:1988**
Escape routes	1 lx minimum There is a UK national exception allowing 0.2 lx in permanently unobstructed escape routes. However, due to the difficulties in keeping escape routes unobstructed during an emergency it is recommended that the 1 lx minimum from BS EN 1838 is always used.	0.2 lx minimum Higher levels were required for routes with obstructions or used by older people but the lighting level is not defined.
Open areas	0.5 lx minimum in core area	1 lx average over total area
Additional areas (e.g. lifts, escalators)	0.5 lx minimum	not specified
High risk task areas	10% of normal illuminance	not specified
All values are designed with zero reflectance		

High risk task area lighting

BS EN 1838/BS 5266-7 says that the average horizontal illuminance on the reference plane (note that this is not necessarily the floor) should be as high as the task demands in areas of high risk. It should not be less than 10% of the normal illuminance, or 15 lx, whichever is the greater. It should be provided within 0.5 s and continue for as long as the hazard exists. This speed of response can normally only be achieved by a tungsten or a permanently illuminated maintained fluorescent lamp source. As the normal lighting level is an average value the emergency value is 10% of that average so calculation is relatively simple. It is even acceptable to provide a conversion kit with a 10% ballast lumen factor or fully run one fitting in ten to achieve the illumination required.

The illumination is only required within the vicinity of the hazard so the required illuminance can often be achieved by careful location of emergency luminaires at the hazard and may not require additional fittings.

Design control procedures

The illuminance of the installation depends as much on the light distribution as it does on the light output available from the chosen luminaire. Consequently, luminaire types specified for a particular design must not be changed without a reappraisal of the photometric design. The completion certificate should identify the fittings used in the photometric design, the installer and verifier must then confirm that they were the fittings used.

Testing and log-book

The system should include adequate facilities for testing and recording its condition. These need to be appropriate for the specific site. Particular care must be taken for the annual full rated discharge test which leaves the system unable to provide its duty again for 24 h until it has been recharged. It may be reasonable to perform a full discharge test of the installation in an office block by isolating the total supply when the premises will be empty. However, this would be inappropriate and potentially dangerous, in a hotel or old people's home that is occupied 24 h a day. A test system able to operate alternate fittings would be more suitable to eliminate the risk of having all the luminaires discharged while the building is occupied.

BS 5266-1:1999 test regime

This regime called for a function test for a short period once a month to check that the luminaire is working.

A discharge duration test is also required. When self-contained fittings are new this test is for one-third of their rated capacity every six months (this hopefully retains some battery capacity if immediately after the test there is a mains failure.) After the fittings are three years old and approaching their four year minimum design battery life the test should be done annually for the full rated duration. The latest edition of BS 5266-1:2005 now aligns with the testing shown in BS EN 50172/BS 5266-8.

A function test for a short period once a month is required to check that the luminaire is working. An annual discharge duration test is required for the full rated discharge.

System design

NOTE: The risks that any tests will materially discharge the battery must be minimized either by ensuring the building will be empty during test and recharge or alternate fittings should be tested.

Test records

A log-book should be provided and be kept readily available for inspection. It should record the date and brief details of completion, any alterations, periodic inspections and test certificates, each service, inspection or test carried out, defects and remedial action. The model test log in BS 5266-1 is shown in Annex A of this book. It reminds the test operative of the safety precautions they should take prior to conducting the test. In addition to recording the results of the tests the log now monitors the rectification of any faults that are found and requires that additional checks are made on the safety of the installation while the repairs are conducted.

The action needed during the repair period depends on the type of risk that is represented. For instance:

- Minor fault: A few luminaires just failing to give the full 3 h illumination. The remedial action would be for a competent engineer to order replacement batteries or luminaires. The safeguards for occupants would be to inform the staff that the emergency period is reduced until replacement batteries are fitted and tested satisfactorily.
- Medium fault: Some luminaires are not functioning but they are not in critical positions. The remedial action would be to introduce additional safety precautions such as issuing staff with torches which will need a weekly test and/or initiate extra patrols until the system is rectified.
- Major fault: The output fuses in a central system are blown in a basement disco. The remedial action would be to organize a competent engineer to rectify the fault and identify and rectify the cause of the system overload. To safeguard the occupants, no one else should be allowed into the building until the engineers have completed the rectification.

As preventive action, adequate maintenance should be actioned and appropriate stocks of likely spares kept onsite. In the case of self-contained luminaires it is often preferable and most cost effective to keep spare luminaires which can then be used to provide a total replacement. The old fittings can then be repaired when convenient and become spares themselves. The time taken to action repairs will be reduced if the current contact references for an appropriate competent engineer and the system manufacturer are kept readily available. Essential servicing should be defined to ensure that the system remains at full operational status at all times. Consumable items such as replacement lamps for maintained luminaires should be available for use as soon as black ending of the service lamps becomes apparent.

Central battery systems should be maintained by a competent engineer. Although many systems now use recombination cells which do not require topping up the connections should still be cleaned and tightened regularly. The charging should be checked to ensure it is within operational limits and each battery block voltage checked to ensure that no cells have become short-circuited. The output circuit also needs to be tested to ensure that the load has not been increased beyond the design values.

The model completion certificate detailed in BS 5266-1:2005 is shown in Annex B of this book. This written declaration of compliance should be available onsite for inspection.

The certificate now comprises an initial page in which the owner/occupier who is responsible for the building should make a declaration that in his or her belief it complies to BS 5266-1:2005. This page must detail the deviations from the standard that have been agreed and it is only valid when accompanied by:

- a signed declaration of design, installation and verification compliance by competent engineers;
- appropriate photometric design data;
- a suitable test log (a model log is shown in Annex A).

Major items covered by the competent engineer's declarations should include.

- Design acceptability confirming that the standard is met and that appropriate areas of the building are protected.
- Product quality. A declaration that compliant equipment was used.
- Installation quality. The wiring installation must conform to the wiring regulations HD 384, and suitable cable, with adequate support and protection, must be used.
- Photometric performance. Evidence of compliance to the design criteria has to be obtained. ICEL 1001 registered fittings are

System design

photometrically tested and their spacing data is registered by the ICEL scheme. Copies of this data provide the verification required so long as the spacing is not exceeded. Alternatively, an appropriate computer print out can be used but it must ensure that appropriate factors have been included to compensate for aging of the luminaires, the effects of dirt and for the reduced output at the end of the discharge period.
- A declaration of a satisfactory commissioning test of operation and compliance to BS 5266 must be made.

These new forms can be used on new or existing installations but specific additional checks are needed on the test records of existing installations.

13. System selection

System requirement

While the major requirement of emergency lighting is to ensure that appropriate safety legislation is complied with, the systems also have other uses and these should be taken into account when selecting the most appropriate system for a particular application.

The major requirement is normally to meet the legal requirement to prevent loss of life or injury when lighting fails, from hazards such as stairs, floor openings, obstructions and high inertia machinery, and to enable personnel to locate then use the escape routes safely. However, it also has a value in preventing financial loss due to theft, accidents, loss of process continuity and loss of working time. These considerations can influence the choice of the most appropriate system.

Examples of these applications that may be important to the user include: a sudden loss of light which creates conditions that encourage 'opportunity theft'. This can determine the best location for the luminaires and the output from them while still fulfilling their safety function. Obvious areas that are at risk are cash in open tills and small size high value goods, these include gambling chips, for example on a roulette table. The appropriate illumination level tends to be higher than for the safety of the staff, but in all cases the speed of the illumination being provided is important. This favours maintained fluorescent luminaires or high power tungsten beam units. In addition, if the lighting failure has no accompanying problems, the operator and the occupants may want to conduct transactions on their way out. For example, a supermarket may wish to have adequate illumination over the checkouts. It will often be preferable for the store manager to check out customers rather than for them to leave their trolleys around the store requiring them to manually return the goods to stock.

Power source

Emergency lighting may be obtained from several types of system.

Electric storage battery systems

These are battery systems charged from and monitoring the mains supply and illuminating lamps in the event of supply failure. They are designed to provide specific finite duration and coverage of emergency lighting and they are the most widely accepted emergency lighting power source.

Advantages of storage battery systems

These provide:

- almost instantaneous illumination availability;
- they are physically compact, so little or no special space is required;
- low cost for low power ranges;
- low maintenance requirement and easy testing;
- it is possible to continuously monitor that the system is healthy.

Disadvantages of storage battery systems

- The finite limited discharge duration has to be followed by a lengthy recharge period.
- Battery cost limitations make them only suitable for providing illumination for safety applications not continuous power provision.

They are of two major types: self-contained where the luminaire contains its own battery and charger which power associated control gear and lamp in the event of a supply failure. Alternatively, a central battery power system can be used where a single large battery and

System selection

charger which in a normal supply failure powers a wired distribution system with a number of slave remote luminaires.

Advantages of self-contained luminaires

- They can be cheaply and quickly installed with minimum disturbance of the decor by using PVC cables in existing buildings.
- They are easy to wire so that they operate when the local lighting fails (this is required for non-maintained fittings).
- Systems can easily be extended by adding more single point units (SPUs).
- The failure of one unit will not render the others inoperative.
- There are no system engineering considerations, such as supply voltage drop calculations.
- They are often the only economic solution for small systems.

Disadvantages of self-contained luminaires

- They require monthly testing.
- They have a short cell life (minimum design life is 4 years but 5–6 years is typical).
- Cell replacement is costly and time consuming.
- The cost of keeping the nickel–cadmium cells fully charged is a high annual expense.
- The light outputs are limited typically to 200 lm by the power available from the built-in batteries. This is adequate for most corridors but not economic for large areas with high ceilings.

Central systems

Advantages of central systems

- They allow control of the system as the maintained circuits can be locally or remotely switched and the output can be inhibited to prevent unwanted discharges when the premises are empty.
- The remote battery is much more cost effective than the batteries used in self-contained luminaires so much higher light outputs can be provided reducing the number of luminaires needing to be installed particularly in open areas with high ceilings.

- The systems offer a choice of battery types, voltage, duration and design life that can be matched to the application.
- Testing is initiated at a single point so that it is more easily carried out.
- They allow a wide choice of luminaires, including using unmodified normal mains luminaires.
- The state of the system can be easily and automatically monitored.

Disadvantages of central systems

- They require a system engineer to design the system.
- They cannot be easily extended, future expansion needs should be built into the initial design.
- They require protected cabling rated to reduce voltage drop problems.

Generator powered systems

A generator produces electricity to operate luminaires, and is powered by petrol, diesel, gas or gas turbine motors. These are generally limited to uses where high power outputs or long emergency lighting durations are required. The long start-up time is often in excess of the 5 s or 15 s which is required for emergency lighting applications. Consequently a separate battery powered system is also necessary to bridge the gap between supply failure and illumination being available from the generator.

Advantages of generator powered systems

- The generators can operate indefinitely, subject only to the supply of fuel, and stoppages due to maintenance and break downs.
- They can use the existing mains fittings.
- They can power a variety of other loads as well as the lighting.
- Very high power outputs may be obtained more cheaply than with any other system, enabling operation of the site to continue, hence providing standby as well as emergency operation.

System selection

Disadvantages of generator powered systems

- There is usually a significant delay between mains failure and the generator reaching operating speed which delays the availability of power requiring battery systems to be used to bridge the time period until the illumination is available.
- The wiring must be in fire resistant cable and each area must have luminaires fed directly from the normal supply while emergency supply from the generator or its by-pass circuit must be connected to the luminaires in the event of a failure of the normal final lighting circuit.
- A separate area is required for the generator and fuel store.
- The stored fuel itself represents a fire hazard.
- The testing schedule required is to run the set monthly for one hour while supplying the majority of the load, after this the set must be refueled to be ready for an emergency duty.
- Noise and vibration cause disturbance and suppression is costly.
- If the generator fails repairs can be lengthy, leaving the premises unprotected.

Typical applications

Generators are only normally used for standby duties where they provide an alternative to the normal mains supply providing full illumination for as long as may be required. In applications like hospitals, essential areas such as operating theatres also need to have battery back-up systems to guard against cabling faults or starting problems with the generator. Other applications use large generators to enable activities to continue with a combination of battery supplied luminaires to bridge the time gap while the generator starts and reaches operating speed and voltage output.

Modes of operation

Maintained systems

The lamps are illuminated at all material times. The luminaires are normally powered from the mains supply, and fed from the battery in the event of normal lighting failure.

Maintained changeover

The lamp is energized from the normal AC supply via a transformer or mains ballast, the battery charger only supplies charging current, and in the event of supply failure a changeover device transfers the load to the battery.

Advantages of a maintained system

- It may be a legislative requirement.
- The continuous illumination of the lamps proves the wiring and the lamp filaments.
- Since the lamps are illuminated at all times it is not necessary to monitor local lighting circuits or phases, only the circuit supplying the maintained lighting.
- In some areas this lighting system may replace mains fittings, saving cost and improving the decor.
- Can be used as a security night watchman's system.

Disadvantages of a maintained system

- The regular switching and burning of the lamps reduces the lamp life so regular maintenance is necessary to inspect for and replace faulty lamps.
- The continuous use of the lamp requires the use of an aging factor to be applied on lamp output photometric calculations.
- The operation of the lamps does not prove the state of charge of the battery.
- The maintained circuit can incur cost and the heat generated reduces battery life in self-contained luminaires.

Combined maintained systems

This form of maintained circuit allows the luminaire to have one or more additional normal mains lamps which significantly increases the illumination available from the luminaire in the mains healthy condition.

System selection

Non-maintained systems

The lamps are only illuminated when the normal lighting fails. A switching device (contactor, relay or solid-state switch) switches the lamp to the battery when the supply fails.

Advantages of non-maintained systems

- The lamps do not age and rarely fail so full design light output is available at end of discharge.
- They normally have lower initial unit costs.
- As there is no maintained circuit heat is minimized and the self-contained battery life is maximized.

Disadvantages of non-maintained systems

- Regular testing is required to prove the load circuit and lamps.
- As fluorescent lamps have to be started they often must be de-rated as they do not reach their optimum output within 5 s and they cannot normally be used in any way for high risk task applications.
- The luminaire only provides emergency lighting so it does not contribute to or form part of the normal lighting system.
- They are unsuitable for use to illuminate signs in locations where the occupants are unfamiliar with the building or in other areas requiring maintained operation.
- Monitoring of phases and local lighting circuits may be necessary, resulting in additional relays and wiring, hence cost, unless a single-point unit system is used.

Combined non-maintained systems

These were previously known as sustained systems.

To avoid running the emergency lighting lamps while still providing a high light output in the mains healthy condition from the luminaire this system uses an extra non-maintained lamp in the housing with a lamp or lamps directly supplied from the mains ballast which is independent of the emergency lighting circuit. This system was popular when tungsten lamps were used as an emergency source as they had a very limited operational life. Now that fluorescent lamps are common the advantages of this configuration are reduced. This system commonly uses a smaller size of emergency non-maintained systems lamp to the normal lamps. Care needs to be taken in testing this system to check that the emergency circuit is providing the illumination.

System choice

The essential elements of any battery powered emergency lighting system are a battery, charger, switching device and lamp. The light sources may be either fluorescent lamps, tungsten filament types or light emitting diodes (LEDs).

Lamp type

Fluorescents are the main lamp of choice as they offer a good efficiency which is typically 60 lm W^{-1}. They also have a reasonably good operating life from 9,000 h upwards depending upon their size. They have to be operated from a controlling ballast providing AC and an appropriate starting circuit but this does allow them to be optimized for output and also enables them to operate from a range of supply voltages.

Conventional tungsten filament lamps have a much lower level of efficiency, normally approximately 10 lm W^{-1}. They do not require any control gear and can operate equally well on AC or DC systems. But their output varies considerably in response to any variations in supply voltage. Their operational life is short unless the lamps are significantly under run with correspondingly large drops in their light output. Remaining applications tend to be applications where the aesthetic quality of the lamp overrides its life and efficiency problems.

Tungsten–halogen lamps offer double the light output of conventional filament lamps. While this is still much lower than for fluorescent lamps, the format of the lamp allows it to be produced with a very small point source of light encouraging its use with appropriate reflectors which can give a tightly controlled beam for use with projector fittings. As they also provide very fast illumination, when connected, they are useful in high risk task applications or for high mounting covering the aisles in between the racking in storage areas.

LEDs are developing rapidly and are producing very encouraging efficiencies but their biggest impact is in offering extended operational life with replacement intervals of four years being currently achievable at good light output. Extensive development is continuing in increasing their power levels to avoid the need for the use of high numbers of individual devices for conventional luminaire use but they are already well established as the light source for exit route signs.

Major system configurations

Self-contained

Early self-contained luminaires were produced containing one or more tungsten filament lamps with a recombination maintenance free battery, a charger and a changeover device (relay or solid-state switch). These luminaires are unsuited to maintained operation unless under-run, when a much lower output is obtained. These units are now mainly used as beam projector units and often use a relatively large lead acid battery which is lower in cost than the equivalent nickel–cadmium type to compensate for the low lamp efficiency.

Self-contained fluorescent units use a solid-state inverter ballast to provide AC to the fluorescent tube from the recombination battery. The tubes are run at high frequencies where the light output efficiency is higher and give a high light output. They typically are under run with outputs of 100 lm from 2-cell circuits and 180 lm from 3-cell circuits.

Maintained versions are available but the circuit to power the lamp in the mains healthy condition incurs cost and generates heat that may reduce battery life. If the maintained circuit needs to be switched off an extra switched supply must be wired to the fitting.

Central systems requiring special luminaires

There are two types of central system requiring special luminaires. First, there are AC/DC wall mounted systems. Compact battery/charger units are available which include a single charger, a changeover device and a recombination battery within a wall mounting cubicle. These are designed for providing local power supply to the emergency luminaires in a specific area, as they are often of 24 V and the proximity of the luminaires reduces the difficulties in providing a satisfactory cabling distribution. Luminaires have to be compatible with the output from the system and the cabling voltage drop.

Secondly, there are AC/DC floor standing systems. These are similar to the wall mounted units except they are available with a wider range of battery capacities, types and system voltages. e.g. up to 150 A h^{-1} and up to 24, 50 and 110 V. A major application of this type of system is to provide immediate local back-up for hospital operating theatre lights. In these applications the effect of voltage drop in the slip rings which allow positioning of the lamp must be taken into account in both the AC (normal) and DC (emergency) conditions.

Central systems able to work with normal unmodified luminaires

AC/AC floor standing systems with central battery units incorporating an inverter which gives an output of 240 V at 50 Hz compatible with normal mains supplies are available. Thus, a wide range of normal mains luminaires may be used for both normal and emergency lighting. Inverters are available from approximately 800 W upwards. Maintained systems may be simply provided by feeding AC mains to the lamps via a bypass connection to the changeover device when the mains are healthy. A variation is available by using a UPS – these systems provide a no break supply. These units must be designed specifically for the application to ensure they are able to clear any distribution fuses or MCBs so that short-circuits that may occur on any part of the distribution can be isolated and the remainder of the system can be restored. In addition, they must be able to start the load from the battery in an emergency and the charger must be adequate to recharge the battery to 80% of capacity within 12 h.

A number of parameters affect the choice between central systems and SPUs, and these are:

Installation

Self-contained luminaires each require to be connected directly to an unswitched mains supply. If the mains supply or the wiring to it fails, the SPU will operate. Thus it is not necessary to use fire resistant cabling. Unless local regulations or specifications state otherwise, PVC insulated PVC sheathed cable of the type used for domestic lighting installations will be adequate. This is easy to lay and to introduce into voids etc. for concealment and has low material and labour costs. It may also be concealed easily with relatively little damage to decor. Self-contained luminaires may also be connected to local lighting mains supplies rather than a central supply which further reduces costs and disturbance to the decor.

System selection

Central systems, both AC and DC, rely on the wiring between the battery and the luminaires for operation. This must therefore be protected against fire. This is achieved by using fire resistant cables unless the construction of the building, having PVC sheathed cables buried in plaster or carried in conduit or trunking, may also be used.

Voltage drop in cables

When a current flows in any wire a small voltage loss will occur which is proportional to the current and related to the size of the wire. Thus the higher the current and the greater the distance travelled the greater the loss.

With self-contained luminaires the current is only flowing in the cables when the mains are healthy and the voltage drop is low since the current is low. When the mains have failed, except when projector fittings are used remote from the power pack, self-contained luminaires suffer no measurable voltage drop problem due to the proximity of the battery and lamp.

Central systems on the other hand, operate at low voltages, and the lamp and the battery are remote. Current is carried by the wiring during emergency conditions. Thus voltage drop represents a power loss, and at low system voltages, voltage drop as a ratio of system voltage becomes of increasing significance. For instance, the light output of a tungsten lamp is reduced by 20% for a voltage reduction of only 5%. Increasing the system voltage will reduce the current and hence the voltage drop, and the higher system voltage further reduces the voltage drop in proportional terms. Increasing the cable size also reduces the voltage drop. Increased system voltage increases the number of cells in the battery, whereas an increase in wire size increases material costs and installation time. These costs must be considered in the choice of central system voltage.

Luminaires

The choice of luminaire appearance is restricted for self-contained luminaires by the limited ranges and performance characteristics. A wide range of luminaires is available which may be adapted for use with DC central systems by adding a fuse and possibly changing the lamp holder. Similarly, a wide range of normal mains voltage luminaires is available for use with AC central systems.

However, the need to design for minimum illumination levels required by modern legislation means that photometric data is needed for emergency lighting luminaires. This should be available for self-contained luminaires and selected central system luminaires. For any other luminaire or for a known luminaire operating at a different voltage, from a different ballast or with a different lamp, further photometric data may be required.

Batteries

With the exception of projector units, most self-contained luminaires incorporate recombination maintenance free batteries. These may be of the nickel–cadmium type with a minimum life of 4–7 years, full performance over the rated life and simple charging requirements, or the cheaper but more bulky lead acid type with a shorter life, a loss of performance with age and more closely controlled charging requirements. Nickel–metal hydride batteries are also being used as they have a slightly better energy density and a physically smaller battery can be used.

A wide range of batteries is available for use with central systems. The principal type now is the recombination lead acid cell, other types being vented lead acid and nickel–cadmium batteries.

Four types of battery are normally used.

- Recombination lead acid offers a cell which is operated without being topped up. When operated with an appropriate charging system, cells can give a life expectancy in excess of 10 years. They have a reasonable initial cost and the removal of the need for access to top them up considerably reduces the size of cubicle they need for accommodation.
- Lead acid plate offers a high performance battery with long life (approximately 25 years) and low maintenance requirements (9-month intervals) but fairly high cost and low capacity to volume ratio. This cell is unique as it retains its capacity throughout its design life as it regenerates active material to replace that lost in use.
- Lead acid flat plate offers a medium performance battery with acceptable life (10–12 years) low cost, good performance and good capacity/volume ratio.
- Nickel–cadmium alkaline offers a long life battery with a good capacity/volume ratio, high resistance to abuse, long periods

System selection

between maintenance, but with a lower voltage (and hence light) at the end of discharge, and high cost.

Maintenance

Self-contained luminary diffusers need cleaning to maintain their optimum photometric performance and ensure that illuminances are met. In addition, maintained lamps should be replaced when they show signs of aging such as blackening of the tube ends.

Central systems with vented batteries require battery maintenance at regular intervals, typically six months.

Load circuit protection

Compliance with the standards requires that failure of a local lighting circuit will bring on emergency lighting. This may be easily achieved by supplying self-contained luminaires from the same fuse. Central systems will either require detection relays or for the system to be maintained.

Extending the system

Self-contained luminaire systems can easily be extended by adding more luminaires. Central systems can only be extended if spare capacity is available. If this is not the case a further system will be required.

Monitoring

Accurate monitoring of system condition, wiring, battery etc. is cheaply carried out with central systems but will be required for each individual unit with SPUs.

Cost

For very small systems, the cost of self-contained luminaires will nearly always be less, but as the system size increases, the cost per lumen or cost per luminaire falls for central systems but remains the same for SPUs. This soon outweighs any additional wiring costs.

Convenience of operation

Self-contained luminaires will readily provide local protection for an area. However, if switchable maintained illumination is needed the central system offers a simpler solution as well as an ability for discharges to be inhibited if required when the building is empty.

Testing

Self-contained luminaires require their supply to be interrupted to operate a test. The device to facilitate this must be able to switch the incoming mains supply for a few minutes every month and for the full rated discharge period annually. Note switching the supply off turns the emergency lamp on, powered from its internal batteries. It must be safe to use, must not switch off any other essential services, and must be adequately labelled. During the test the operation of the luminaire must be checked and recorded then the mains restored and the luminaire charging lamps checked to ensure they are operational again. If the building is always occupied the annual duration test should be conducted in such a manner as to limit the risks of an emergency occurring just after the test with the luminaires all discharged. To ensure that each compartment of the building has some light alternate luminaires should be tested.

The same tests are required to be performed on central battery systems. If the system provides a maintained output the connection to the luminaires and their operation can be checked from the maintained supply, thus reducing the time for which the battery is operated.

If the building is continuously occupied duration testing cannot be achieved by checking alternate luminaires, but fortunately the characteristic of the industrial batteries enables them to be duration tested for two-thirds of capacity. During that time the battery output voltage must not have fallen below a preset value (normally 5% higher than the normal full duration end of discharge value).

Either type of system can benefit from the use of automatic testing systems as defined in IEC 62034.

14. Photometry for emergency lighting

Photometric theory

Propagation of light/basic photometric units

We see an object or a surface because light is reflected back from it to the eye. The incident light arrives after flowing through space from the light source. It is convenient to think in terms of four stages:

- light source;
- flow to the surface being illuminated;
- arrival at the surface;
- return from the surface to the eye.

Numerical interpretation is possible at each stage. We speak of the intensity of the source, measured in candelas; the luminous flux measured in lumens; the illuminance, measured in lux (lumens per square metre); and finally, the brightness of the surface, or its 'luminance', measured in candelas per square metre ($cd\ m^{-2}$).

Intensity/standard candle/candela

In the early attempts to measure light it seemed reasonable to speak of the intensity of a source in terms of how it compared with the commonest everyday source of artificial light, the candle. The modern unit of luminous intensity, the candela, is the result of the search for the greatest possible precision in defining a standard source.

Luminous flux/intensity and flux

For most practical lighting design what matters is the total luminous flux emerging from a source, the total flow of light into the complete space around it. The simplest case is for a point source which radiates light uniformly in all directions around it. The lumen can be defined as the flow of light through an area of 1 m^2 on the surface of a sphere of 1 m radius with a uniform point source of 1 cd at its centre. The total number of lumens emitted in all the space around this source is equal to the number of times an area of 1 m^2 might be fitted into the complete surface of the 1 m radius sphere.

Illuminance

The illuminance of a surface is a measure of the concentration of light falling on it. To express it in numbers we need the amount of light, the lumen (lm), so that we can say that the output of a lamp is a specific number of lumens, or that the required illuminance for a particular location is so many lumens per unit area.

Lux

Illuminance is measured in lumens per square metre or 'lux' (lx). This is the value that is used to define emergency lighting requirements.

Basic principles

Two factors reduce light levels as measurements are taken further away from the luminaire. The inverse square law states that: as the light moves away from the source the angle of the light beam reduces to a specific area, so consequently the light falling on that area reduces by the inverse square of the distance.

The cosine correction factor states that: as the measuring point moves away from having been directly under the luminaire, the reducing angle of incidence to the floor decreases the amount of light from each angle of output beam hitting a specific area. This reduction in light is given by the cosine of the angle to the luminaire.

Photometry for emergency lighting

Lighting requirements

BS EN 1838/BS 5266-7 defines that the escape routes and open area measurements are the worst case that will be encountered in the life of the system so the light intensities must be calculated with de-rating factors applied for all variables. Because the emergency lighting will not be modified if the décor changes, the system has to provide the minimum required light levels at all times in the system's life.

This means that it must cover the reduced light output likely to be provided as the battery discharges and during the lower light that occurs from a fluorescent lamp as it warms up during starting. Calculations have to account for the aging of lamps and other components during their life and the effects of dirt reducing the light output between cleanings.

Also the illumination has to be normalized to that falling at 90° to the floor which is considered as the working plane. This then allows the calculated light from two or more luminaires to be added, for example in an open area in between four fittings, if 0.125 lx is provided by each fitting then the requirement of 0.5 lx will be met.

Point intensity calculation

To save complicated calculations these two factors are resolved by the use of a single combined formula.

The light level at a specific point in lux is $$\frac{I \times \cos^3 Y}{H^2}$$

Where I is the intensity of the light source, Y is the angle of incidence and H is the height of the luminaire above the floor (see Figure 15). This calculation enables the designer to define the light level at a point provided that they know:

- the intensity from the luminaire at the angle Y;
- the height that the luminaire is mounted above the floor (the working plane);
- the horizontal distance of that point from the luminaire.

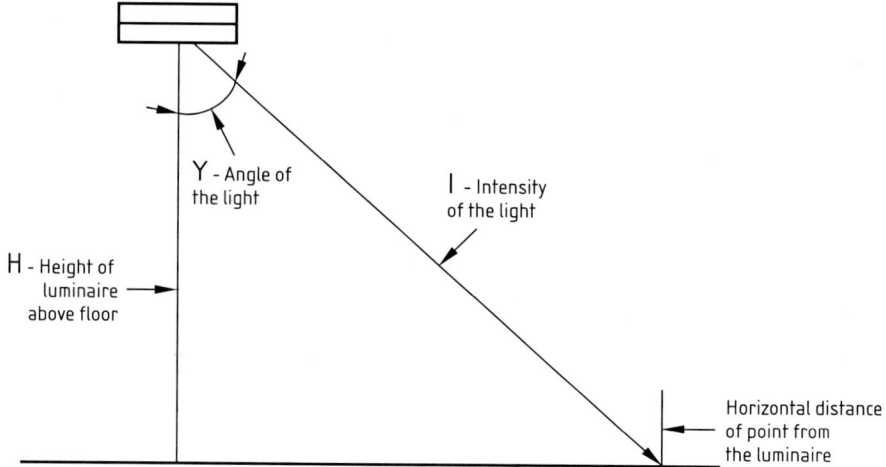

Figure 15 Point intensity calculation

Manual calculation of spacing tables

While there are times when this value is useful, normally we need to know the distance from the luminaire at which a required illuminance is given.

To determine this distance a number of points need to be calculated and a curve drawn to give the precise value. It is normal practice to construct a table and use the tables for cos3 values given in the charts in Annex D of this book.

This process obviously takes a considerable time which is why authenticated spacing tables are produced for most dedicated emergency fittings. However, to be sure they have been calculated correctly it is sensible to only use authenticated tables that have been third-party verified.

The ICEL Scheme 1001 was set up to ensure that valid data is issued by manufacturers. To be registered in this scheme the luminaires have to be fully tested to BS EN 60598-2-22 and the photometric data from that test is checked by independent engineers. This ensures that the derived tables are accurate and use appropriate de-rating factors to compensate for the effects of dirt, the reduction in output of the battery during discharge and the aging of the components.

Spacing tables show, for selected mounting heights, the distances that the first fitting should be located from the start of the escape route and then the maximum distance to subsequent fittings to maintain the minimum level of illuminance on the centre line of the escape route. It is important that the orientation of the fittings is correctly observed. The spacing table defines the distances in both axial and transverse directions. These are calculated from the appropriate C0 and C90 plane basic data.

Tables are also constructed for optimum spacing in open areas to provide a minimum of 0.5 lx anywhere on the central core. Because a room is unlikely to fit the spacing exactly, experience allows these tables to be used with discretion, for example if the number of transverse mounted fittings works out at 6.1 it is often possible to use 6 instead of rounding up to 7 if the axial spacing has already been rounded up and will compensate for the difference.

The success of this scheme has resulted in it being used as a basis for the draft CEN document which will enable data from any compliant country to be used consistently anywhere in Europe.

Computer design

Spacing tables are easy to use but by their nature they provide data for specific conditions. Greater accuracy can be obtained by using computer programs which can be used for precise applications. The programs can be used either with dedicated emergency luminaires or for normal mains luminaires powered either fully or partially from an emergency circuit.

Specific conditions that can be identified include.

- The precise mounting height (spacing tables typically show only 2.5, 4 or 6 m heights and values in between have to be interpolated).
- Angles of orientation (luminaires may not be fixed on a horizontal surface or they may not be mounted at 0° or 90° to the escape route).
- Higher levels of dirt allowance (spacing tables are normally calculated with an 80% allowance for dirt but this may not be appropriate for the site).
- Particular luminaire and circuit combinations (there are vast potential options of luminaire distribution and emergency output that can be used for emergency lighting and it would not be practical to produce spacing tables for each option).
- The isolux chart produced by the computer identifies the light level for 50% of corridor widths.

Use of computer design

It is feasible to enter the format of each room and then add the fittings to provide either the isolux plots or tables of actual values. This may well be the best procedure for open areas. But it is often preferable to use the computer to generate custom spacing tables which can then be used in the normal way to engineer escape routes and so avoid the need to draw up complicated corridors.

To produce a spacing table it is possible to place a single fitting in a large room to produce the calculations, but it is easier to locate the luminaire in the point of intersection of two arms of an L-shaped room each arm 2 m wide and with measurement areas placed for half the corridor width. Then each measuring area provides all the data needed to produce the tables. In either case the room should be calculated with the walls designed for zero reflectance.

If the room is chosen to have a height greater than any fittings that will actually be mounted, then dropping down to the actual heights can be done without redrawing the room or re-entering the fitting.

Verification of photometric design

To be able to demonstrate that the emergency lighting system is adequate, it is important that the light level is correctly designed initially, and then the final system must be able to be verified as having the correct illumination level.

The installation can be verified using authenticated photometric design data which enable systems to be checked for compliance with BS 5266-1 and BS 5266-7/BS EN 1838 without resorting to onsite light level tests.

Compliance can either be demonstrated by ensuring that the design is within the maximum spacing allowed by third-party authenticated spacing tables for the appropriate light level. Or, by use of an appropriate computer design package which uses BSI or other nationally approved test house measurements of luminaire output and distribution and calculates the conditions for the use of either spacing tables or computer design data.

- The data can only apply to control gear which has been photometrically tested by BSI or an equivalent test house and to luminaires which are built under ISO controls to ensure that all the production meets the performance of the tested unit.

Photometry for emergency lighting

- The computer program and input data must meet all the de-rating requirements of the standard to ensure that the minimum design lighting levels are achieved and maintained throughout the system's life.

BS 5266-1 and BS 5266-7 require that authenticated photometric design data must be supplied to demonstrate the conformity of the installation with the design requirements and specifically that it complies with BS EN 1838.

Authenticated photometric design data has to be available for inspectors as a part of the completion certificate. It can be in two forms. Either, dedicated emergency luminaires spacing tables produced and registered as authenticated data by ICEL (these can be photocopied directly or the data from them can be reproduced but they must relate to the precise model used). Or, specific computer developed data can be produced using the same factors and principles for BSI tested control gear by using a program which uses established basic point to point calculation techniques. The computer program and input data must meet all the de-rating requirements of the standard to ensure that the minimum design lighting levels are achieved and maintained throughout the system's life.

It is very difficult to measure the installation for the low light levels that we need to measure. Considerable care is needed to black out any stray incoming light even when the test is performed at night. Acceptable readings have to be taken after 5 s and 60 s and at the end of the discharge. These readings need to be taken at all points that could give a minimum result.

Once the readings have been taken they have to be assessed to ensure there is a sufficient safety factor so that reductions in performance because of aging, the effect of dirt and changes of décor will not reduce the light levels below the standard requirements during the life of the system. Overall, this is not an ideal procedure as it is prone to giving distorted readings, is time consuming, has to be conducted at unsocial times and requires an expensive instrument to achieve reasonable accuracy.

Products that are difficult to provide with photometric data

Chandeliers and other fittings with cut glass faceted optics can often be often be required to be used as the emergency luminaire if they are also the normal lighting luminaire. Generally, an overall calculation should be made assuming that the fitting acts as a bare lamp with a safety margin added for the effects of the optics. After instillation it may be necessary to take confirming measurements if there are doubts about the adequacy of the illumination.

Adjustable angle beam projector units tend to use tungsten halogen light sources and can be useful in some applications such as down the aisles of racking. However, they have a surprisingly narrow beam typically of only 60° so six projectors would be needed to cover all the angles in the centre of an open area.

Only a few manufacturers have had their projectors photometrically measured and have then produced correct escape route spacing data. Care should be taken to ensure that the spacing tables cover the points under the luminaire as well as at the centre of the spacing between projectors and that information should also be given of the aiming angle at which the beams should be mounted.

The problems of satisfactorily engineering the photometric design of this product and the low efficiency of its light source compared to fluorescent fittings are reducing its use. But it can still be used to cover high risk task areas as the illumination is quickly available and the beam enables the light to be concentrated at a specific hazard.

15. Design considerations for major applications

Different types of premises have their own specific problems that will affect the design and selection of the most suitable type of system.

Hospitals and nursing homes

The need for and the suitability and adequacy of the fire precautions will rest with the health authority and the fire authority. It is therefore essential that there should be early discussion and consultation involving all interested parties.

Particular fire hazards

Gases often have to be used and some need to be readily available. They contribute to the fire hazard and need to be stored correctly.

Special circumstances

Persons using these premises will almost certainly be physically incapacitated to a greater or lesser extent, and/or could be mentally impaired. Because of this, and because those undergoing treatment as in-patients (as well as visitors) will be unfamiliar with the overall layout, all these premises require special consideration. When dealing with such premises, it is essential to have regard to all of these factors and look carefully at problems that may arise because of some of the physical requirements of the residents/patients.

The following circumstances will have to be catered for: patients who are confined to bed and would need a considerable evacuation time; wheelchair patients who may need staff assistance; patients using walking frames; patients needing physical assistance (e.g. blind or deaf); and convalescent homes for children.

Because of the likely condition of patients, hospitals do not normally evacuate fully. Patients are moved from the risk and the fire is fought floor by floor. Consequently great care is taken to limit flammable materials in these buildings.

The duration of the system also changes in concept because of the need to continue working in as much as possible of the hospital. Hence it is normal to use generators as an alternative to the normal supply with battery supplied luminaires protecting against local distribution failures.

The levels of illumination should be sufficient to allow movement to be easy, particularly in those areas where there are elderly patients.

Where critical tasks are being carried out, e.g. in operating theatres, the system must cater for immediate lighting by independent battery supply until the standby generator is operating.

Emergency lighting

The majority of the users of these premises are likely to be suffering from some physical defect (and in elderly person's homes, defective vision is likely to be prevalent).

Consequently the emergency lighting may need to be at a higher level than would be necessary in other types of premises.

Standby lighting will be required in hospitals to enable certain essential activities to be carried out. In an emergency hospital authorities normally work to two standards of illuminance. In critical areas, such as operating theatres, delivery rooms and high dependency units, the quality of standby lighting should be equal or nearly equal to the normal lighting. Consideration can be given to the use of portable lighting fittings for tasks outside the critical working areas.

The two standards of standby lighting recommended are.

- Grade (a) lighting of intensity and quality equal or nearly equal to that provided by the normal lighting.
- Grade (b) a reduced standard of lighting, e.g. about one-third to one-half the normal standard, sufficient to enable general hospital activities to be properly carried out.

Design considerations for major applications

NOTE: In hospitals and nursing homes, the overall light output from the normal supply is frequently dimmed by selective switching or dimming of lamps. At such times, this lower lighting level is interpreted as 'normal lighting' and the standby lighting, if activated during this period, whether grade (a) or grade (b) as defined above, should relate to the light output under that dimmed condition.

Emergency escape lighting

Because of the high ratio of infirm occupants who may require assistance, a minimum level of 1 lx is recommended on the centre line of all escape routes. In hospitals, safety lighting will generally be supplied from automatic start standby generating plants. A short break in supply can occur with this type of system during the engine run-up period, unless supplemented by a back-up battery source on automatic changeover, giving a one-hour duration. If this supplementary emergency lighting is not provided and a break in supply is unacceptable for certain areas, e.g. staircases, then battery operated luminaires should be fitted.

Types of luminaire and mounting height

Cleanliness is obviously of high importance and luminaires should be easy to clean and without dust traps. If luminaires are used within the line of sight of a patient's bed head, care must be taken to reduce the glare to acceptable limits.

Operating theatre lights

These have a particular importance and problems. The provision of emergency lighting here is of the highest importance and it needs to be of the same level as normal lighting. This is normally achieved by a maintained central system of 24 V DC. Points for consideration are that: the operating lamp supply is normally passed through slip rings, which drop 0.75 V and typically, luminaire currents are in the order of 40–50 A. So to overcome the effects of supply voltage drop to the lamp, both maintained transformer output and battery should supply 26 V of full load. Also, the maintained transformer should only supply a single main luminaire as load switching would vary its output voltage and the corresponding light output.

Staff

Staff accommodation is often located remotely from the main buildings and would normally be protected by its own battery systems. Staff may need to move in and out of the premises during the hours of darkness, and there may also be security issues that should be considered.

Testing procedures

Testing needs careful consideration as these buildings are always occupied so the procedure will have to be able to be conducted without disconnecting any other essential service and at minimum risk should a supply failure occur during the duration test or while the batteries are being recharged. This would normally be achieved by testing alternate fittings.

Associated equipment: fire alarms

Early detection of a fire is a high priority but this is normally only identified to staff. Most hospitals have a direct connection to their fire brigade.

Mains lighting

Because of the need to reduce the fire loading of these buildings to a minimum, self-extinguishing housings of the mains luminaires are required. As these sites are normally designed for a long operating life, robust construction of luminaires is required.

Hotels and boarding houses

Special circumstances

As these premises are sleeping risks, emergency lighting should be provided in all these buildings identified by the fire safety risk assessment. It must be capable of illuminating all stairways, exit routes, exit and directional signs sufficiently to enable persons to make their way out of the premises. Occupants of these premises may be aged or

Design considerations for major applications

infirm and only resident on a short-term basis, and they may therefore, be totally unfamiliar with the layout of the premises. Residents may also be suffering from the effects of alcohol and confused as to the location of exits.

Luminaires and signs must therefore be positioned to show clearly the exit routes and final exits from premises, changes in floor level (e.g. stairs), changes in direction (particularly on stairs) and corridor intersections.

Premises with lifts need considerable care as users may never have used the emergency stairs. So signage needs to be clear to show the location of the stairs and repeated at any points where the direction of the route may be ambiguous. This also serves to reassure users that they are progressing correctly.

Emergency lighting must be provided in public areas such as dining rooms, ballrooms etc., to enable progress to an escape route to be made round existing hazards, such as tables etc. The lighting level in these areas may, therefore, need to be higher than along a clearly defined escape route. The emergency lighting must be kept on at all times when the rooms are in use and when there is insufficient natural light for escape purposes.

Emergency lighting

In dining halls and other public areas 0.5 lx may be appropriate even for rooms smaller than 60 m^2 but which have an escape route passing through them. In staff working areas such as kitchens, which come under the Health and Safety at Work requirements, additional hazards could be caused by cooking appliances etc. There may be also be obstructions and difficulty in obtaining clear access to an escape route.

Usually, the appearance of the luminaires in public areas is of importance. The luminaires therefore need to be compatible with the decor and styling of the building, and where appropriate able to be controlled to be compatible with different functions.

Types of luminaire and mounting height

Where low levels of illumination are needed at night maintained emergency luminaires can be employed instead of using the high output mains fittings.

In kitchen areas etc., enclosed luminaires should be used, which can easily be kept clean and are not adversely affected by steam and high temperatures. Locating fire extinguishers in any area that is used for cooking requires emergency lighting coverage.

The use of internally lit directional signs is to be preferred, even though externally lit signs are permitted. An internally lit sign is more conspicuous than a painted sign. This is particularly important to identify stairs which residents would not normally use.

Associated equipment: fire alarms

These premises require a high level of protection from the fire alarm system as they contain a large number of sleeping residents. Also there are increased risks from the fact that some smoking materials may have been inadequately extinguished and will re-ignite later.

Mains lighting

In public areas aesthetic considerations are important to ensure that the luminaires match the décor and create a suitable ambiance. Because the building is in operation for 24 h of each day, the efficiency of the luminaires chosen is important in order to reduce the energy consumption.

Non-residential premises used for recreation

Special circumstances

This includes such premises as theatres, cinemas, concert halls, exhibition halls, sport halls, sports stadia, public houses and restaurants. Many of these premises will not be evacuated in a mains failure, so a minimum of 3 h duration should be provided.

The people using such premises may be expected to be unfamiliar with their layouts and exit signs should be maintained to maximize their conspicuity. In parts of premises where the normal lighting may be dimmed, a maintained emergency lighting system should be installed to ensure that minimum rate levels of illumination are met.

For some cinema and theatre auditoria where the recommended maintained illuminance is likely to affect normal working, it may be considered acceptable to reduce this level to one that is safe for normal operation with minimum movement of occupants. In this case the system should be arranged so that in event of failure of the normal system of lighting within the auditoria, the escape lighting illuminance is immediately and automatically restored to a higher level to enable emergency evacuation to be conducted quickly.

Restaurants may normally use candles for illumination and ambience. This will increase the fire risk of the premises. The normal mains lighting may be healthy but switched off making the use of maintained emergency fittings necessary.

Emergency lighting

Complete or substantially complete blackouts, which may be required for production reasons, may only be permitted with the approval of the enforcing authority.

In other places where it is desired to reduce the artificial lighting for effects purposes (e.g. a discotheque), it may be permissible with the approval of the enforcing authority, to extinguish the maintained emergency lighting, provided that the switching for this arrangement is under continuous management control and that the area is visible from the switching position. It is essential that the circuit and equipment adopted is such that the emergency lighting is automatically restored in the event of the failure of the normal supply. Exit signs shall remain illuminated at times when the premises are occupied.

Sports arenas and stadia may involve hazardous areas for competition, in which case they shall be treated as high risk task areas to the values and durations given in BS EN 12193 which details an interim lighting level dependent on the normal lighting to enable the participants of the sports to safely acclimatize to the lower level and cease their activities when the normal emergency lighting becomes adequate.

In many of these premises, occupants are not evacuated in the event of a normal supply failure. In these cases, extended durations are required for premises with a central system. It is useful to provide an indication of approaching end of duration to initiate evacuation while system capacity is still available.

Generators are sometimes used in larger buildings, but if this method is to be acceptable for escape lighting, then the required illumination must be available within 5 s, otherwise panic can occur. For this reason, it is also common practice to provide a battery-operated back-up system to cover the start-up period of the generators.

Types of luminaire and mounting height

The choice of luminaire and the mounting height are closely related. High brightness luminaires, such as floodlights or large fluorescent luminaires, may need to be mounted high to avoid glare and to obtain the maximum spacing within the 40:1 uniformity ratio required. However, care must be taken to ensure that they are not in danger of being obscured by smoke, which may rise to the higher parts of the building.

Luminaires of lower brightness are also suitable for high mounting, but should preferably be positioned at the 2.5–3 m level, if the construction of the building permits. For ease of maintenance, selection of luminaire type should be considered for the application, for example appropriate appearance, lack of glare in operation and adequate physical strength for sports areas.

Associated equipment: fire alarms

As many of these premises are licensed by local authorities they have to meet their requirements as well as normal standards

Mains lighting

The mains lighting should be related to the task being performed, in cinemas and theatres a low level of illumination is needed during the performances which is integrated into the emergency system with higher levels available during the intervals as management lighting. Sports stadia, however, need much higher levels of illumination in most areas, particularly if television broadcasts are likely from the venue.

Shops and covered shopping precincts

Special circumstances

The requirements for emergency lighting in shops are as varied as the shops themselves, ranging from single-roomed high street shops, through multistorey departmental stores to hypermarkets. The main objectives of escape lighting are to indicate and illuminate the exits and escape routes, and to illuminate hazards such as displays, trolleys, wire

Design considerations for major applications

baskets, cartons etc., to protect the customers, but staff areas like stock rooms and rest rooms must not be overlooked.

Additional emergency lighting for security purposes at cash desks and areas where expensive goods are displayed is worth consideration, as is standby lighting, which would enable business to be carried on in the event of a power failure. Particular attention should be given to supermarket cash desks, which often can cause restricted exit routes that need high-risk protection.

Covered shopping precincts also vary widely in size and the layout can have several levels.

Emergency lighting

Circulation areas are normally the responsibility of the precinct management, whereas the shops etc. will be the responsibility of each individual shop management. Common areas need to have an emergency duration at least equivalent to any premises they serve.

If the circulation areas are not adequately illuminated, people could be moving from shops which have emergency lighting, into dark, crowded areas. This could cause panic and it is therefore essential to provide:

- adequate illumination for general movement;
- good directional indications.

Stairs, moving stairways and walkways (possibly still moving), passageways and exit ramps are all potential hazards and should be well illuminated. Car parks should also be provided with adequate emergency lighting as people can be expected to converge on these areas in any emergency situation.

Types of luminaire and mounting height

The choice of luminaire and the mounting height are closely related. High brightness luminaires such as floodlights or large fluorescent luminaires, need to be mounted high to avoid glare and to obtain the maximum spacing. Care must be taken to ensure, however, that they are not in danger of being obscured by smoke, which may rise to the higher parts of the buildings. If the building ceiling has a smoke reservoir, the emergency luminaires must be located below it and be of maintained type.

Associated equipment: fire alarms

Each shop unit should be interconnected to the common areas so that a fire in one shop is indicated to other occupants of the building.

Mains lighting

Different types of lighting are used to produce the most attractive display of goods or provisions varying from clear bright illumination from tungsten halogens, for example on fish counters, to a softer more discrete illumination for other applications.

NOTE: In the vicinity of cold cabinets luminaires should provide a minimum of heat output as this increases the size of cooler needed.

General industrial premises and warehouses

Special circumstances

Since industrial premises vary so widely, there is no simple solution to satisfy the various and often conflicting conditions. Generally, the premises tend to be large with hazardous routes from the workplace to an exit. For example, many warehouses have tall, narrow passageways between shelves or stacks of pallets, and although the stacking areas are often clearly defined, the stacking heights will frequently vary as stocks come and go. Exit points may often be hidden from direct view and the escape routes need to be clearly defined.

In factories the passageway onto escape routes from production areas could be extremely hazardous when the normal lighting fails. This will be particularly so if the power supply for the lighting is separate from that to machinery, tools, conveyor belts etc., as these could continue to operate. It has also to be recognized that even if the power supply to all machinery is interrupted simultaneously with the power supply for the lighting, certain processes are unlikely to come to an immediate halt. For example, rotating machines may take time to come to a complete stop, and chemical and heating processes in ovens or vats will continue. Low luminance contrasts between obstructions and their background, narrow walkways etc., may require an emergency lighting illuminance level higher than that necessary for a simple escape route.

Gatehouses should also have emergency lighting as they are often the designated control points in an emergency and are likely to be the area where the fire alarm control panel or a repeater panel are located.

Emergency lighting

It will be necessary, however, to provide a higher level of illuminance in the potentially dangerous areas leading onto escape routes and to allow for difficulty in movements in areas where emergency shut-down procedures have to be carried out. The level of illuminance needs to be adequate to permit those procedures to be completed safely. This is usually 10% of the normal lighting level.

Types of luminaire and mounting height

Locating luminaires too low renders them liable to damage from fork-lift trucks etc.

In installations using a small number of high output luminaires, special effort is required to keep within the 40:1 uniformity ratio and a maximum spacing to height ratio of 4:1 is recommended. If high pressure lamps provide the normal lighting, provision should be made to cover their re-strike after a short supply break. This is normally by the use of either maintained emergency luminaires, or artificially extending the period of emergency duration to cover the output until the normal luminaires light.

Associated equipment: fire alarms

Care has to be taken to ensure that the forms of detection offered are appropriate to the risk and will not cause false alarms by responding to any of the industrial process being conducted.

Mains lighting

Areas with low ceilings may be treated in the same way as offices or hotels with regard to the type of luminaires but in areas with high ceilings the choice is more difficult. High pressure discharge luminaires are frequently used. These cover a wide area from a small number of positions with the advantage of reduced installation and maintenance cost. However, care must be exercised in positioning them to avoid glare, without putting them in the apex of the roof where they could be obscured by smoke.

Offices

Special circumstances

Offices even if designed to be open areas tend to become compartmented as staff create work stations so emergency escape routes need to be kept under review. For offices which are smaller than 60 m^2 and only need emergency lighting because an exit route passes through that area provision of 0.5 lx in the core area will accommodate any changes in layout. If some staff work late they are often then in isolated locations and the risk assessment will be likely to identify the need to ensure that the areas they need to pass through in their exit are illuminated.

Emergency lighting

As with most other types of emergency lighting, the main task will be to define the escape route. In the case of a cellular office complex, this may be self-evident. Passageways which lead towards doors will be fairly obvious and the lighting layout may be considered as being where a multiplicity of small areas, private and general offices lead into main escape routes. Thus the basic rules will apply: luminaires will be at intersections and changes in direction or level etc.

With very large or open-plan offices, however, the problem becomes more difficult. The furniture layout may be irregular and an escape route may be difficult to define and keep clear. Also the layout of some offices often changes and hence the position of the exit routes. Any point in the room must therefore be regarded as part of the escape route. Exits should be indicated by means of signs visible from every point in the room.

Open plan offices larger than 60 m^2 should be provided at a minimum of 0.5 lx anywhere. This will allow any new layout to be used as required without relocating the luminaires.

Design considerations for major applications

Types of luminaire and mounting height

The choice of luminaires and mounting height are closely related. The competition for scarce ceiling space in modern offices can lead to the inclusion of emergency gear within the luminaires (conversion units). This creates problems if the normal luminaires are of the low brightness type for use with visual display terminals (VDTs), which have louvres 'cutting off' lighting distribution. In these cases either a conventional distribution non-maintained luminaire should be used or a larger number of the normal fittings will be needed and should be used with low output emergency circuits to meet the uniformity requirements.

Associated equipment: fire alarms

The normal considerations apply here.

Mains lighting

Offices now are normally illuminated by luminaires with louvered diffusers which cut off light to reduce the veiled reflection of the lamp in computer screens. Work stations should be assessed to ensure that they conform to the Visual Display Terminals Regulations.

Schools and colleges

Special circumstances

In general, the layout and safety provisions of this type of premises, are usually familiar to those persons using them and orderly evacuation can normally be expected in the case of an emergency. Evening activities are, however, now common at such places and the emergency lighting system should be planned with knowledge of the extent to which such activities are likely to occur.

Where there are boarding facilities, then the requirements for hotels should be used as a guide. Unless natural light can be guaranteed at all material times, emergency lighting should also be provided in the work areas, as for normal industrial premises. Again, the minimum lighting level provided will very much depend upon the circumstances and a higher level of emergency lighting than normal may have to be provided where there are handicapped students.

Although the high level of arson suffered by schools tends to take place outside normal school hours, the increasing use of the buildings for community activities increases the chances of people being present when the arson occurs. Alarm systems need to balance the needs for safety with site security. These sites are at high risk from arson.

Emergency lighting

No special requirements are necessary unless the school is for physically or mentally handicapped students, where special provisions may have to be made, particularly if the school has residential units.

In addition to coverage of escape routes such as corridors and stairs, normally the assembly hall and gymnasia will be larger than 60 m^2 and should be treated as open areas. If used for more than occasional use (typically four times per year), they will need to be treated as open areas.

Also, although a classroom may be less than 60 m^2, the risk assessment may identify hazards such as a chemistry laboratory with glass jars of acid that could be dropped in darkness, thus demonstrating a need for emergency lighting to open area requirements.

Kitchens and workshops (with woodworking machines etc.) may also require special attention because of the additional hazards likely to be present and the possibility of obstructions and difficulty in obtaining clear access to an escape route.

Types of luminaire and mounting height

No special requirements are necessary apart from congregational areas such as gymnasia and assembly halls where the exit points must be clearly identified, preferably by internally illuminated maintained signs, as visitors may be present.

Luminaires used in gymnasia and in other places of sport and recreation should have the light-emitting diffusers protected by wire guards, or be constructed of suitable impact resistant material.

Associated equipment: fire alarms

Systems may be required to indicate class change in addition to alarm signals. Break glasses may be fitted with a clear cover to reduce 'accidental' activation. Traditionally the alarm bells were also used as a signal for class changes. Now that electronic sounders with different signal options are available, this possible confusion can be avoided.

Mains lighting

Rugged luminaires are needed in this application. Particularly in gymnasia, it may be desirable to use either impact resistant polycarbonate fittings or wire guards can be used to provide protection. Good exterior lighting is also valuable to reduce the risks of arson and vandalism that schools suffer from.

Transport locations

These areas cover: airports, railway stations, underground stations and bus depots. These places have particular problems such the presence of a high number of people, often encumbered by luggage, who may be unfamiliar with the premises.

Special circumstances

By the nature of their operation, all ages and a mixture of people may be present and can cause serious problems when trying to evacuate the buildings, particularly as language differences may make understanding of verbal instructions difficult.

By nature of the forms of travel, the sudden arrival of a passenger plane or train instantly deposits a high number of people in the building who must be handled safely.

Failure of the lighting supply to a train station or airport may not be allowed to stop the arrival of further planes or trains so the emergency plan has to be able to cope with these additional passengers during the emergency period.

- Airports. There is likely to be a large number of people unable to understand verbal instructions, so clear signage is very important. In addition, security and passport controls may need to be able to evacuate occupants to segregated secure areas.
- Railway stations. Ticket barriers generate restricted escape routes that need high levels of illumination and can have very high volumes of people in restricted areas of platforms. The platforms themselves are a potential high risk in darkness.
- Underground stations. In addition to normal railway station hazards, below ground evacuation is more difficult and long tunnels can create dangerous fire chimneys. If lifts are normally used, in mains failure evacuation by stairs will be slower and the users will be unfamiliar with the route and need good signage.
- Bus depots. High numbers of people may be present in an area used by vehicles and taxis. Protection of illuminance is needed to reduce accident risks at the moment of normal supply failure.

Because of the high profile of transport areas and the difficulties protecting them, they have been subject to terrorist attacks and vandalism. Adequate reliable illumination reduces their vulnerability. If closed circuit television is an important part of the security and crowd control, the level of emergency lighting should be adequate to enable surveillance to be maintained.

Emergency lighting

Because of the high risk in many transport areas, higher levels of emergency lighting and effective signage is needed in many areas such as platforms, ticket barriers, changes of level etc. Because of security consideration and the fact that many travellers do not have anywhere else to go, extended duration is normally required.

Types of luminaire and mounting height

For many cases, weatherproof fittings are needed and vandal resistance should also be considered, both by the strength of the luminaires and by its mounting height and location.

Associated equipment: fire alarms

The system has to be able to be heard above any other noises. If voice alarm signals are used it may be necessary to also use different languages.

Mains lighting

Typically high mounting heights favour the use of discharge lamps as the light source but care should be taken to ensure people are safe during re-strike conditions.

16. Installation, maintenance and testing of emergency lighting

Initial procedures

When considering installation it is important to ensure that the correct equipment has been supplied and that it has not suffered any damage in transit. The manufacturer's instructions should be consulted to check for any special action that may be needed.

Self-contained systems

Self-contained luminaires are generally simple to install provided simple rules are followed.

When the correct luminaire has been obtained for the location a final check should be made to see there are no special site conditions that need attention, such as the need for a weatherproof or a vandal resistant fitting.

The luminaires need adequate fixing as, with their self-contained batteries, many fittings are fairly heavy, particularly if they are to be mounted on a suspended ceiling.

Special cabling is not normally required as any interruption of the supply will activate the luminaire. However, as non-maintained luminaires sense this supply it is important that they are connected to the final circuit of the lighting for the mains luminaires in the area. This restriction does not apply to maintained luminaires which will be energized at all times that the building is occupied. To ensure this the supply to any maintained circuits that are to be switched should be ganged with an essential circuit such as the mains lights.

Consideration must be given to the testing procedure that will be used; if the building is never unoccupied it is advisable to be able to test alternate luminaires so that the building is never left with compartments without any light. The test devices should be appropriate to test for a few seconds for the monthly function test but also for the rated duration annually. Automatic test systems need to be set for a safe period to conduct the annual test.

After installation the supply should be conducted and operation of the charge healthy indicator checked. Provided all is well the units should be allowed to charge for at least 24 h and then a full rated discharge should be performed as a commissioning check. The results should be checked and entered in the test log. If any luminaires do not reach their full rated duty the charge discharge cycle should be repeated to see if the battery cycling has restored the required duration.

Maintenance

As with all luminaires the diffusers of emergency units should be cleaned to restore photometric performance. Maintained lamps will also need to be replaced to compensate for aging during service.

Rectification of faults

The charger indicator monitors that the mains supply, charger output and battery continuity are healthy. Fluorescent lamps tend to blacken giving warning of impending failure and should be replaced. If the duration is reduced this indicates that the battery needs replacing. This should be done with the correct battery type as the cell life is matched to the application. This is particularly true of conversions of mains luminaires that are run as maintained luminaires as high temperature performance needs to be matched to the application.

If excessive battery consumption is detected the charge rate should also be checked but this can only be done after 24 h on charge as the taper charger output does not stabilize until the cell is fully charged.

Central battery systems

The location of the central power supply unit should be in an area of low fire risk and having normal ventilation. The manufacturer's instructions should be carefully followed. Particular care needs to be taken with batteries as their stored energy, even of low voltage blocks, is considerable and at high voltages special precautions need to be taken. Cabling should also be routed through low risk areas where possible and fire protected cables and joints should be used. Any breaches of the fire compartment during installation should be made good.

The luminaires are installed in the same way as normal luminaires but in compound fittings which have both normal mains supply and the emergency circuit in the luminaire adequate insulation and safety warnings must be fitted. Control relays switching mains AC and the output of inverters need to be able to withstand both out of phase voltages.

Care should be taken regarding the number of luminaires protected by a distribution fuse or protective device. The latest guidance is that 20 fittings with a maximum consumption of 6 A is an appropriate limit per protective device.

Testing procedures need to ensure that any hold-off relays are activated to obtain the full load on test. If dimming circuits are present on any of the luminaires they must also automatically revert to full output in a supply failure or in the event of a fire.

The luminaire load from a single central battery cannot be tested alternately, so if the building will not be empty during the duration test the system should be tested for two-thirds of the duration. The system manufacturer will advise precise values, but at the end of two-thirds of the discharge the battery should typically be above 90% of nominal battery voltage. (In full discharge the system voltage will be at least 85% of its nominal battery voltage).

During initial commissioning the battery should be fully recharged for 24 h then the test should be applied and the battery voltages at two-thirds of duration and full duration recorded so that future partial discharge tests can be validated.

The testing records become an essential part of the completion documentation needed to hand the system over.

When carrying out maintenance, in addition to luminaire cleaning and lamp replacement, the battery terminals should also be checked for cleanliness and tightness.

Rectification of battery faults likely to be identified in routine testing

If the battery follows its normal voltage reduction curve but is consistently 2 or 4 V low it is likely that one or two cells are shorted out and a specialist battery engineer will be able to rectify the system. If, however, the battery is failing to meet its duration period it is likely that a replacement will be needed.

Although central systems are now fitted with much more detailed system condition monitors, the user needs to understand the information that they are provided with and also when they should call in specialist help to maintain the systems.

For all systems it is sensible that appropriate spares are kept available and that the source of replacement essential items such as circuit boards and batteries is known.

Training of the user's staff should be conducted to ensure that they understand the operation and testing procedures for the system. If this is not possible, adequate instructions should be provided to them. All test and isolating switches must be clearly marked and of an appropriate type for their location.

Annexes

The following Annexes are included to help with the design and control of the operation of emergency lighting systems.

Annex A Inspection and test certificates

These forms record the results of the periodic tests on the emergency lighting system. They detail the fault action plan to rectify any equipment failures and the procedures to maintain the safety of the premises while the repair is being made.

These records are required to be able to be shown to inspecting authorities to demonstrate that correct testing has been performed on the installation.

Annex B Emergency lighting completion certificates

These certificates form a declaration that the system conforms to the current emergency lighting standards. The covering sheet defines any agreed deviations from the standards and should contain signed declarations of compliance from the engineers responsible for the system design, installation and verification.

It must be accompanied by photometric design data, and a test record as shown in Annex A for the system.

Annex C Compliance checklist for inspecting engineers

The checklist provides a ready reference list of the major requirements for compliance to enable engineers to audit existing installations. The list is useful for inspection authorities and those engineers modifying or taking responsibility for an installed system.

Annex D Mathematical table for use in photometric calculations

This is intended to assist engineers who may be making a manual calculation or check of photometric design calculations.

Annex E Legislation and standards affecting emergency lighting

A list of the titles of major legislation, standards and guidance documents relating to emergency lighting.

Annexes

Annex A Inspection and test certificates

Emergency lighting inspection and test certificate For systems designed to BS 5266-1 and BS EN 50172/BS 5266-8			
WARNING Full duration tests involve discharging the batteries, so the emergency lighting system will not be fully functional until the batteries have had time to recharge. For this reason, always carry out testing at times of minimal risk, or only test alternate luminaires at any one time.			
System manufacturer Contact phone number			
System installer Contact phone number			
Competent engineer responsible for tests		Phone number	
Site address			
Responsible person			
Date the system was commissioned			
Details of system mode of operation	Non-maintained		
	Non-maintained luminaires, maintained signs		
	Maintained		
	Other		
Duration of system	........hours	Is automatic test system fitted?	Yes/No
Details of additions or modifications to the system or the premises since original installation			
Addition or modification		**Date**	
Action to be taken on finding a failure • The supplier of the system or a competent engineer should be contacted to rectify the fault. • A risk assessment of the failure should be conducted. This should evaluate the people who will be at increased risk and the level of that risk. Based on this data and, if necessary, advice from the fire authority, the appropriate action should be taken. *Continued...*			

A Guide to Emergency Lighting

- Action may be:
  - to warn occupants to be extra vigilant until the system is rectified;
  - to initiate extra safety patrols;
  - to issue torches as a temporary measure;
  - in a high risk situation, to limit use of all or part of the building.

NOTE: Test programs for identifying early failures can reduce the chances of failure of two adjacent luminaires at the same time.

Annexes

Emergency lighting inspection and test record			Sheet number	
Site				
Test types: C = commissioning test M = monthly test (see BS EN 50172:2004, 7.2.3) A = annual test (see BS EN 50172:2004, 7.2.4)				
Date of test	Test type	Result: test passed, no action needed	Result: test failed, details of actions needed	
			Repair of system	Safeguarding of premises
	C			
	M -1st			
	M -2nd			
	M -3rd			
	M -4th			
	M -5th			
	M -6th			
	M -7th			
	M -8th			
	M -9th			
	M -10th			
	M -11th			
	A			
	M -1st			
	M -2nd			
	M -3rd			
	M -4th			
	M -5th			
	M -6th			
	M -7th			
	M -8th			
	M -9th			
	M -10th			
	M -11th			
	A			
	M -1st			
	M -2nd			
	M -3rd			
	M -4th			
	M -5th			
	M -6th			
	M -7th			
	M -8th			
	M -9th			
	M -10th			
	M -11th			
	A			

Emergency lighting fault action plan			Sheet number:		
Contact references		Contact name	Phone number		
Equipment supplier:				For replacement parts	
Maintenance engineer:				Competent engineer	
Responsible person:				Site control	
Date of failure	Action taken to safeguard the premises		Action taken to rectify the system		Date system repaired

Annexes

Annex B Completion certificates

For new installations and verification of existing installations

Occupier/owner

....

Address of premises

....

Declaration of conformity

In consequence of acceptance of the appended declarations, I/we* hereby declare that the emergency lighting system installation, or part thereof, at the above premises conforms, to the best of my/our* knowledge and belief, to the appropriate recommendations and requirements given in BS 5266-1:2005, Emergency lighting: Part 1: Code of practice for the emergency lighting of premises, BS EN 1838:1999/BS 5266-7:1999 Lighting applications: Emergency lighting and BS EN 50172:2004/ BS 5266-8:2004, Emergency escape lighting systems, as set out in the accompanying declarations, except as stated below/overleaf.

*Delete as appropriate

Signed, on behalf of owner/occupier

....

Name ....

Deviations from standards

Declaration (design, installation or verification)	Requirement number	Details of deviation

A Guide to Emergency Lighting

This certificate is only valid when accompanied by current:

a) Signed declaration(s) of design, installation and verification, as applicable (see overleaf).
b) Photometric design data. This can be in any of the following formats, but in all cases appropriate de-rating factors must be used and identified to meet worst case requirements.
   - Authenticated spacing data such as ICEL 1001 registered tables**.
   - Calculations as detailed in CIBSE/SLLGuide LG12***.
   - Appropriate computer print of results.
c) Test log-book.

**Available from Industry Committee for Emergency Lighting, ICEL, Westminster Tower, 3, Albert Embankment, London SE1 7SL.
***Available from Chartered Institution of Building Services Engineers, Delta House, 222 Balham High Road, London SW12 9BS, UK.

Annexes

Design: Declaration of conformity

BS 5266-1 clause reference	Requirements	System conforms (if NO, record a deviation)		
		YES	NO	N/A
4.2	**D1** Accurate plans available showing escape routes, fire alarm control panel, call points and fire extinguishers			
6.6	**D2** Fire safety signs in accordance with BS 5499-4 and BS 5499-5, clearly visible and adequately illuminated			
7.9	**D5** The luminaires conform to BS EN 60598-2-22			
6.3.2	**D6** Luminaires located at following positions: (NOTE: Near means within 2 m horizontally) • at each exit door intended to be used in an emergency; • near stairs so each tread receives direct light, and near any other level change; • near mandatory emergency exits and safety signs; • at each change of direction and at intersections of corridors; • outside and near to each final exit; • near each first aid post; • near fire fighting equipment and call points.			
7.4	**D7** At least two luminaires illuminating each compartment of the escape route			
7.8.3 7.8.4 7.8.5 7.8.6 7.8.7	**D8** Additional emergency lighting provided where needed to illuminate: • lift cars; • moving stairways and walkways; • toilet facilities larger than 8 m^2 floor area or without borrowed light, and those for disabled use; • motor generator, control and plant-rooms; • covered car parks.			
10.1	**D10** Design duration adequate for the application			
11.6; 12.3	**D11** Operation and maintenance instructions and a suitable log-book produced for retention and use by the building occupier			

BS 5266-1 clause reference	Requirements	System conforms (if NO, record a deviation)		
		YES	NO	N/A
6.3.2	**D12** Illuminance. Escape routes for any use: 1 lx min. on the centre line Open areas above 60 m^2: 0.5 lx min. anywhere in the core area Permanently unobstructed routes: 0.2 lx min. on centre line			
	NOTE: If the system was designed to the illuminance levels given in BS 5266-1:1988, this should be recorded as a deviation.			

Deviations from standards

(to be entered on completion certificate)

Requirement number	Details of deviation

Signature of person making design conformity declaration ....

Qualifications ....

For and on behalf of ....

Date ....

Annexes

Installation: Declaration of conformity

BS 5266-1 clause reference	Requirements	System conforms (if NO, record a deviation)		
		YES	NO	N/A
6.3	**IN1** The system installed conforms to the agreed design			
7.2	**IN2** All non-maintained luminaires fed or controlled by the final circuit supply of their local normal mains lighting			
7.5	**IN3** Luminaires mounted at least 2 m above the floor			
7.5	**IN4** Luminaires mounted at a suitable height to avoid being located in smoke reservoirs or other likely area of smoke accumulation			
6.6	**IN5** Fire safety signs in accordance with BS 5499-4 and BS 5499-5, clearly visible and adequately illuminated			
9.2	**IN6** The wiring of central power systems has adequate fire protection and is appropriately sized			
9.3.5	**IN7** Output voltage range of the central power system is compatible with the supply voltage range of the luminaires, taking into account supply cable voltage drop			
7.9	**IN8** No glow starters in any emergency circuits			
9.2.13	**IN9** Any plugs and sockets protected against unauthorized use			
9.3.3	**IN10** The system has suitable and appropriate testing facilities for the specific site			
12.1	**IN11** The equipment manufacturer's installation and commissioning procedures satisfactorily completed			
9.1	**IN12** The system conforms to BS 7671			

Deviations from standards

(to be entered on completion certificate)

Requirement number	Details of deviation

Signature of person making design conformity declaration ....

Qualifications ....

For and on behalf of ....

Date ....

Annexes

Verification: Declaration of conformity

BS 5266-1 clause reference	Requirements	System conforms (if NO, record a deviation)		
		YES	NO	N/A
4.2	**V1** Plans available and correct			
9.3.3	**V2** System has a suitable test facility for the application			
6.6	**V3** All escape route safety signs and fire fighting equipment location signs visible with the normal lighting extinguished			
6.3	**V4** Luminaires correctly positioned and oriented as shown on the plans			
7.9	**V5** Luminaires conform to BS EN 60598-2-22			
7.9	**V6** Luminaires have an appropriate category of protection against ingress of moisture or foreign bodies for their location as specified in the system design			
12.2	**V7** Luminaires tested and found to operate for their full rated duration			
12.2	**V8** Under test conditions, adequate illumination provided for safe movement on the escape route and the open areas NOTE: This can be checked by visual inspection and checking that the illumination from the luminaires is not obscured and that minimum design spacings have been met.			
12.2	**V9** After test, the charging indicators operate correctly			
9.2	**V10** Fire protection of central wiring systems satisfactory			
9.2.6	**V11** Emergency circuits correctly segregated from other supplies			
11.6; 12.3	**V12** Operation and maintenance instructions together with a suitable log-book showing a satisfactory commissioning test provided for retention and use by the building occupier			
Clause 13	**V13** Building occupier and their staff trained on suitable maintenance, testing and operating procedures, or a suitable maintenance contract agreed			

BS 5266-1 clause reference	Requirements	System conforms (if NO, record a deviation)		
		YES	NO	N/A
Additional requirements for verification of an existing installation				
Clause 12	**V14** Test records in the log-book complete and satisfactory			
Clause 13	**V15** Luminaires clean and undamaged with lamps in good condition			
6.3	**V16** Original design still valid NOTE: If the original design is not available this needs to be recorded as a deviation.			

Deviations from standards

(to be entered on completion certificate)

Requirement number	Details of deviation

Signature of person making design conformity declaration ....

Qualifications ....

For and on behalf of ....

Date ....

Annexes

Annex C Compliance checklist form inspection engineers

Emergency lighting installation to BS 5266-1:2005				
Compliance checklist for inspection engineers Issue 2 6-7-2005				
Site address		Date		
Responsible person				
No.	Checks including those conducted during work in progress	Y	N	N/A
1	**Check that the appropriate system has been installed and documented**			
1.1	Are the correct areas of the premises covered?			
1.2	Is the system documentation correct and available?			
1.3	Has the system been designed for the correct mode of operation category?			
1.4	Has the system been designed for the correct emergency duration period?			
1.5	Is a completion certificate available with photometric design data?			
1.6	Is a test log available and are the entries up to date?			
2	**Check of the system installed**			
2.1	Are the luminaires installed those documented in the design?			
2.2	Are the exit signs and arrow directions correct?			
2.3	Are there luminaires sited at the 'points of emphasis'?			
2.4	Is the spacing between luminaires compliant to spacing tables or drawing?			
2.5	Is there illumination from at least two luminaires in each compartment?			
2.6	Are the luminaire housings suitable for their location?			
2.7	Are non-maintained luminaires monitoring the local lighting circuit?			
3	**Check of the quality of the system**			
3.1	Do the luminaires conform to BS EN 60598-2-22?			
3.2	If a central power supply unit is used does it conform to BS EN 50171?			
3.3	For centrally powered systems is the wiring fire resistant?			
3.4	Do any converted luminaires conform to BS EN 60598-2-22/ICEL 1004?			

4	**Test facilities**			
4.1	Do the test facilities simulate a supply failure?			
4.2	Are the test facilities safe to operate and do not isolate a required service?			
4.3	Are the test facilities clearly marked with their function?			
4.4	Are the user's staff trained and able to operate them and record correctly?			
4.5	If an automatic test system is installed does it conform to IEC 62034?			
5	**Central powered systems**			
5.1	Are escape lighting components and cables installed correctly?			
5.2	Can any AC systems start the lamps from the battery in an emergency?			
5.3	Can any AC systems blow all distribution fuses/MCBs in an emergency?			
6	**Final acceptance to be conducted at completion.**			
6.1	Are the areas of coverage in accordance with the requirements imposed under the Building Regulations and the risk assessment?			
6.2	For central systems: has the correct cable type been installed?			
6.3	Does the number and distribution of fittings appear to be reasonable?			
6.4	Have escape lighting cables been segregated from all other cables?			
6.5	Is the standard of cable installation satisfactory?			
6.6	Are all isolators, switches and protective devices minimized and marked?			
6.7	Have suitable test facilities been installed and marked?			
6.8	Have all escape lighting cable penetrations been fire stopped?			
6.9	Does the system operate correctly when tested?			
6.10	Has adequate documentation been provided to the user?			
Results of the inspection		Signed		Date
Comments				

Annexes

Annex D Mathematical table for use in photometric calculations

Angle	Radians	Cos	Cos 2	Cos 3	Sin	Tan
0	0	1	1	1	0.0	0
1	0.017	1.00	1.00	1	0.017	0.017
2	0.035	1.00	1.00	1	0.035	0.035
3	0.052	1.00	1.00	1	0.052	0.052
4	0.070	1.00	1.00	0.99	0.070	0.070
5	0.087	1.00	0.99	0.99	0.087	0.087
6	0.105	0.99	0.99	0.98	0.10	0.11
7	0.122	0.99	0.99	0.98	0.12	0.12
8	0.140	0.99	0.98	0.97	0.14	0.14
9	0.157	0.99	0.98	0.96	0.16	0.16
10	0.175	0.98	0.97	0.96	0.17	0.18
12	0.209	0.98	0.96	0.94	0.21	0.21
14	0.244	0.97	0.94	0.91	0.24	0.25
16	0.279	0.96	0.92	0.89	0.28	0.29
18	0.314	0.95	0.90	0.86	0.31	0.32
20	0.349	0.94	0.88	0.83	0.34	0.36
22	0.384	0.93	0.86	0.80	0.37	0.40
24	0.419	0.91	0.83	0.76	0.41	0.45
26	0.454	0.90	0.81	0.73	0.44	0.49
28	0.489	0.88	0.78	0.69	0.47	0.53
30	0.524	0.87	0.75	0.65	0.50	0.58
32	0.559	0.85	0.72	0.61	0.53	0.62
34	0.593	0.83	0.69	0.57	0.56	0.67
36	0.628	0.81	0.65	0.53	0.59	0.73
38	0.663	0.79	0.62	0.49	0.62	0.78
40	0.698	0.77	0.59	0.45	0.64	0.84
42	0.733	0.74	0.55	0.41	0.67	0.90
44	0.768	0.72	0.52	0.37	0.69	0.97
45	0.785	0.71	0.50	0.35	0.71	1.00
46	0.803	0.69	0.48	0.34	0.72	1.04
48	0.838	0.67	0.45	0.30	0.74	1.11
50	0.873	0.64	0.41	0.27	0.77	1.19
52	0.908	0.62	0.38	0.23	0.79	1.28
54	0.942	0.59	0.35	0.20	0.81	1.38
56	0.977	0.56	0.31	0.17	0.83	1.48
58	1.012	0.53	0.28	0.15	0.85	1.60
60	1.047	0.50	0.25	0.13	0.87	1.73
62	1.082	0.47	0.22	0.10	0.88	1.88
64	1.117	0.44	0.19	0.08	0.90	2.05
66	1.152	0.41	0.17	0.07	0.91	2.25
68	1.187	0.37	0.14	0.05	0.93	2.48
70	1.222	0.34	0.12	0.04	0.94	2.75

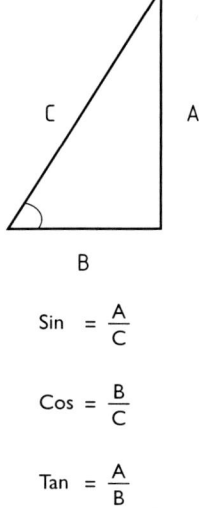

$$\text{Sin} = \frac{A}{C}$$

$$\text{Cos} = \frac{B}{C}$$

$$\text{Tan} = \frac{A}{B}$$

A Guide to Emergency Lighting

Angle	Radians	Cos	Cos 2	Cos 3	Sin	Tan
72	1.257	0.31	0.10	0.03	0.95	3.08
74	1.292	0.28	0.08	0.02	0.96	3.49
76	1.326	0.24	0.06	0.01	0.97	4.01
78	1.361	0.21	0.04	0.01	0.98	4.70
80	1.396	0.17	0.03	0.01	0.98	5.67
81	1.414	0.16	0.02	0	0.99	6.31
82	1.431	0.14	0.02	0	0.99	7.12
83	1.449	0.12	0.01	0	0.99	8.14
84	1.466	0.10	0.01	0	0.99	9.51
85	1.484	0.09	0.01	0	1	11.4
86	1.501	0.07	0	0	1	14.3
87	1.518	0.05	0	0	1	19.1
88	1.536	0.03	0	0	1	28.6
89	1.553	0.02	0	0	1	57.3
90	1.571	0	0	0	1	

Annexes

Annex E Legislation, standards and training affecting emergency lighting

European directives and recommendations

Workplace Directive (89/654 EEC)
Construction Products Directive (89/106 EEC)
Safety Signs Directive (92/58 EEC)
Fire Safety in Hotels Recommendation: Requirements for Europe (86/666 EEC)

UK legislation

Regulatory Reform (Fire Safety) Order 2005 (This replaces the Fire Precautions Act 1971.)
Health and Safety at Work etc. Act 1974
Health and Safety (Safety Signs and Signals) Regulations 1996, Statutory Instrument 1996/341
Safety Signs and Signals. The Health and Safety (Safety Signs and Signals) Regulations 1996. Guidance on Regulations, L64. HSE Books, 1996
Workplace (Health, Safety and Welfare) Regulations 1992 Statutory Instrument 1992/3004
The Building Regulations 1991: Approved Document B Fire Safety and the Building Regulations 1991 Approved Document B Fire Safety Amendments 2002 Appendix B,
Cinematograph (Safety) Regulations Statutory Instrument 1955 No. 1129
Disability Discrimination Act 1995 The Stationery Office 1995

British Standards: General Series and Codes of Practice

BS 5266-1:2005 Code of practice for the emergency lighting of premises (This now includes cinemas and replaces CP 1007:1955 Maintained lighting for cinemas.)
BS EN 1838/BS 5266-7:1999 Lighting applications: emergency lighting
BS EN 50172/ BS 5266-8:2004 Emergency escape lighting systems
BS EN 60598-2-22:1998 Luminaires — Particular requirements for emergency lighting
BS EN 50171:1999 Centrally powered systems

A Guide to Emergency Lighting

BS 5499/4:2000 Code of practice for escape route signing
BS EN 12193:1999 light and lighting: Sports lighting

Draft European and international standards

IEC 62034 Automatic test system for battery powered emergency escape lighting
Emergency escape lighting systems measurement and presentation of photometric data for lamps and luminaires (doc. CEN/TC 169 WG3)

Industry guides

ICEL 1001 Emergency lighting luminaires
ICEL 1004 Conversion of normal luminaires for emergency use

Training

The new legislation moves from using a set of prescriptive safety standards for a given application to requiring the use of risk assessments which may need individual designs to provide the hazard protection needed.

To design their emergency lighting and fire alarm systems employers are recommended to use competent engineers to assist them to decide on the appropriate system.

To ensure that system suppliers and designers are aware of the latest standards and equipment and their engineers are capable of producing appropriate schemes BSI and ICEL have produced a joint training initiative to certificate Competent Engineers. The Competent Engineer Course is intended to support emergency lighting engineers and enable them to apply the new fire safety legislation.

The course is designed to increase the knowledge of emergency lighting engineers to enable them to support and apply the new fire safety legislation it consists of a number of assessed sections covering:

- Legislation – how to apply the information from a risk assessment.
- Relevant standards – with an update of the latest European and International documents.
- Design procedures – how to produce and demonstrate the compliance of designs.

Annexes

- System selection – guidance on the features of particular system types.
- Testing and maintenance – to ensure that procedures are adequate and appropriate.

For further information regarding this course, please contact: seminars@bsi-global.com

Background – Changes in legislation and standards

Currently Fire Certificates are issued by fire authorities after they have determined the level of protection needed and inspected the installed system. They will largely be replaced by placing the responsibility for the safety of premises on the employers who will be expected to follow guidelines to determine the level of risk and to ensure that occupants are adequately protected.

The new legislation relies on employers having information to demonstrate and prove that their premises are safe, this is a considerable change from when the burden of proof was on the fire officers demonstrating that the premises were unsafe.

The scope of the legislation requires all premises employing staff to be checked with a risk assessment. Most existing fire certificates will no longer be valid and premises issued with them to old standards or with equipment that no longer operates properly will need to meet current standards and rectify any equipment faults

This change of responsibility will have a major impact on the motivation of the users. In the past their priority was often just to satisfy the fire officer's initial inspection, now they need equipment which will meet the performance requirements and continue to do so for as long as possible, so components need to be available to maintain its operation.

Employers should use competent engineers to design schemes with the recommended relevant light levels and product standards to protect against hazards identified by their risk assessment.